ROUTLEDGE LIBRARY EDITIONS:
ECONOMIC GEOGRAPHY

Volume 5

GLOBAL CITIES

GLOBAL CITIES
Post-Imperialism and the Internationalization of London

ANTHONY D. KING

Routledge
Taylor & Francis Group

LONDON AND NEW YORK

First published in 1990

This edition first published in 2015
by Routledge
2 Park Square, Milton Park, Abingdon, Oxon, OX14 4RN

and by Routledge
711 Third Avenue, New York, NY 10017

Routledge is an imprint of the Taylor & Francis Group, an informa business

© 1990 Anthony D. King

British Library Cataloguing in Publication Data
A catalogue record for this book is available from the British Library

ISBN: 978-1-138-85764-3 (Set)
eISBN: 978-1-315-71580-3 (Set)
ISBN: 978-1-138-88535-6 (Volume 5)
eISBN: 978-1-315-71549-0 (Volume 5)
Pb ISBN: 978-1-138-88536-3 (Volume 5)

Publisher's Note
The publisher has gone to great lengths to ensure the quality of this reprint but points out that some imperfections in the original copies may be apparent.

Disclaimer
The publisher has made every effort to trace copyright holders and would welcome correspondence from those they have been unable to trace.

GLOBAL CITIES

*Post-Imperialism
and the
Internationalization of London*

ANTHONY D. KING

ROUTLEDGE
London and New York

First published 1990
by Routledge
11 New Fetter Lane, London EC4P 4EE
29 West 35th Street, New York, NY 10001

© 1990 Anthony D. King

Phototypeset in 10pt Baskerville by
Mews Photosetting, Beckenham, Kent
Printed and bound in Great Britain by
Biddles Ltd, Guildford and King's Lynn

British Library Cataloguing in Publication Data

King, Anthony D. (Anthony Douglas), *1931–*
Global cities: post-imperialism and the
internationalization of London.
1. Cities. Development. Socioeconomic
aspects
I. Title
307.'14

Library of Congress Cataloging-in-Publication Data
King, Anthony D.
Global cities: post-imperialism and the internation-
ilization of London / Anthony D. King.
p.   cm.
Bibliography: p.
Includes indexes.
1. Urbanization — Economic aspects. 2. Urbanization —
Economic
aspects — England — London. 3. London (England) —
Economic conditions. 4. Architecture and Society 5. Sociology,
Urban. I. Title.
HT361.K56   1989
307.7'6—dc 19   88-32172

ISBN 0-415-00883-2

CONTENTS

FOR DAVID
14 years' seminars
in the World City

TABLES

PREFACE

This book originated as a case study of London's role as 'world city', the first version of which was completed at the close of 1983 as part of a ten-city project on 'World cities in formation', organized by Professor John Friedmann, then Head of the Urban Planning Program, University of California, Los Angeles. Drawing on the research agenda suggested by Friedmann and Wolff (1982), the aim of the project was to explore the link between urbanization processes and global economic forces.

That not all the case studies were completed and published turned out, as far as the London project was concerned, to be fortuitous. The economic forces that had already made such a massive impact on, for example, Los Angeles and New York City (well described in the various publications of Soja, Sassen-Koob, and others), though powerfully present in London, were still building up in the early 1980s; and the consciousness of, and literature on, the internationalization of capital and its effects were still relatively limited.

The full impact of these forces was only to be realized with impending deregulation in the City in the mid-1980s and 'Big Bang' in 1986, since when, the amount of published literature has enormously increased. With the vast quantity of both journalism and academic writing published in the last five years, it is difficult to believe that a case had to be made (in the UK at least) for research on the relation of cities to the world-economy (King, 1983; Thrift, 1985).

Yet in the six years since Friedmann and Wolff's article was published, spelling out some of the major changes taking place in world cities, its main arguments have become increasingly relevant for the UK as a whole and London in particular. And despite recent critiques (Korff, 1987), Friedmann's 'world-city hypothesis' fulfils its objectives as 'a framework for research and . . . a starting point for political enquiry' (Friedmann, 1986: 86). It is a revised version of this case study, therefore, selectively updated to the end of 1988, that forms the second, and major half of this book.

However, two other developments have taken place since 1983. The first is the rapid growth of literature on world-city formation and of the impact of world economic forces on urbanization. A brief overview of this is provided in Chapter 2. Yet, in reviewing this literature, it became apparent that, in focusing as it generally did on the impact of world economic forces on cities in the 'advanced industrial' countries at the core of the world-economy, it had largely overlooked the experience of other cities, equally affected by the development of the world-economy but removed, both in time and space, from those that were currently the focus of interest. These were the earlier colonial cities situated on the global periphery.

Despite the fact that colonialism and imperialism (two sides of the same coin) were instrumental in creating the present world-economy, their significance for the development of the modern urban system has been largely ignored. Hence, rethinking and developing earlier work on colonialism and colonial cities, and relating it to the emergence of the world-economy (*Urbanism, Colonialism, and the World-Economy* (King, 1989a)) became a necessary pre-requisite for understanding the recent development of 'world' or 'global' cities.

As indicated in that volume, the modern colonial city, whether in South Asia, Africa, South-East Asia, Australasia, or in the 'informal' or 'post'-colonial economies of Latin America, provided the historically first site for the confrontation and interaction of capitalist and pre-capitalist societies, for their respective economic, social, and political institutions, 'developed' and 'underdeveloped' economies, and for representatives of widely different races, ethnicities, religions, and cultures, on any major scale.

Yet it was precisely because of the hegemonic power of the formal and informal imperial states that this interaction of the metropolitan and colonized, of Europe and non-Europe, was confined to the colonial cities. It is a major thesis of this book (and hence the subtitle) that a significant factor contributing to the emergence of world cities today, whether in the core or peripheral regions of the world-economy, is the disappearance of this hegemonic control of formal empires. Whether in allowing the freer flow of capital, goods, services, or labour, the dissolution of empire has been critical to the growth of world cities.

It is because of this inherent connection between colonialism and the world-economy, and between colonial cities and world cities, that the first part of this book, building on the arguments spelt out in *Urbanism, Colonialism, and the World-Economy* (King, 1989a) puts together work on colonial cities and world cities, together with theoretical formulations on both. Its objective is to contribute towards a more satisfactory, theoretically and historically informed understanding of both the similarities and differences in today's global cities.

The aforementioned comments indicate the historical context in which the book developed. Some reference is also needed to the geographical and social context.

My interest in some of these issues arose from approaching the understanding of urbanization in Britain from the 'outside in' rather than the 'inside out', from the experience, some years earlier, of examining urbanization in India. In subsequent years, my perspective has also been formed by teaching on various graduate courses in different university institutions in London (indicated in the acknowledgements) with students from different parts of the world though with perhaps rather more from 'ex-colonial' Commonwealth countries, and generally, though not always, market economies. By comparison, fewer students have been from Eastern or Western Europe and the USA.

The book, therefore, comes from attempts to develop a framework and conceptual vocabulary with which to discuss different aspects of cities, the production of the built environment and the forces that influence it to students that might typically come from Mexico, Ecuador, Nigeria, Canada, Greece, India, Trinidad, Turkey, Kenya, Tanzania, Hong Kong, Malaysia, Singapore, Algeria, Iran, and elsewhere. The following chapters, therefore, reflect this situation as also, a particular view of 'global cities' as seen from London — just as one might say that other recent work on the world city reflects a view from Los Angeles or New York. And in commuting weekly between Yorkshire (Leeds) and London for half of the year between 1974 and 1987, I have also gained an insight into (and demonstrated) both the increasing spatial reach (200 miles) of London as well as urban restructuring in the North. Since January 1988, the commuting has been extended to the USA.

ACKNOWLEDGEMENTS

I am particularly indebted to four people whose help has been instrumental in the preparation of this book.

My first thanks are due to John Friedmann of the Urban Planning Programme at the University of California, Los Angeles, whose invitation, in 1982, to participate in a larger project on world-city formation prompted my interest in the phenomenon and resulted in an earlier version of the London case study included here. Whilst I have strayed from the earlier framework of this project, I am none the less grateful for his many comments during the first stage of that research.

At the Development Planning Unit of the Bartlett School of Architecture and Planning, University College, London, Mike Safier's enthusiasm, knowledge, and critical comments in our discussions over the years have also been much appreciated as also have his comments on earlier drafts of some of these chapters. I am very happy to acknowledge his help.

Third, though her inputs into the general theme of the book have been from a different perspective and of a different nature, Ursula King, of the University of Leeds, has made innumerable contributions, both visible and invisible, intellectual and practical, to the finished product. I owe her much more than these few words of acknowledgement can express. And on a more domestic level, I would like to thank both her, as also Frances, Karen, Anna, and Nina, as always, for their real, varied, and practical support.

My greatest thanks, however, are reserved for David Page, whose unstinting kindness and generous hospitality in London over many years have provided the context for the discussion of countless issues, not only in this but in my previous books. Whether in this, or in contributing to my understanding of many aspects of the world economic and political situation, David's generosity has been boundless (not least in the tolerance extended to my different viewpoints) and it is to him, as 'a small token of my appreciation', that the book is dedicated.

In addition, both David Simon (Royal Holloway and Bedford College, University of London) and Nigel Thrift (University of Bristol) have offered valuable comments on some earlier drafts; I am additionally grateful to Nigel for allowing me to pillage from his many papers.

Many of the ideas in the book have been developed in the context of different graduate programmes with which I have been associated in the last few years, which have encouraged me to think about the production of the built environment at different levels: that of urbanization and the city (urban development planning), urban sector (urban design), building (building design), and housing (architecture). At the Development Planning Unit, University College, London, in addition to Mike Safier, my thanks are due to Ronnie Ramirez, Nigel Harris, and other colleagues; at the Joint Centre for Urban Design, Oxford Polytechnic, Ivor Samuels and his colleagues have provided an invaluable opportunity to discuss the political economy of urban design; reading student theses from all over the world (including one-time colonies) has prompted many of the questions that are addressed here; at the Graduate School of the Architectural Association School of Architecture, London, Jorge Fiori, Clare Johnson, and colleagues have provided an open (and especially flexible) environment in which to discuss the production of dwelling form in developing countries and finally, in the new course on Building Design in Developing Countries at the Bartlett School of Planning and Architecture, Cho Padamsee and his students have prompted many of the reflections about the influence of colonialism and global capitalism on building and urban form. And in encouraging the prompt completion of the book, the State University of New York at Binghamton (and my colleagues in the Art History program, whom I also wish to thank) has reinforced one of its main messages.

Outside university walls, Ian Robinson of Brunel, the University of West London, has, for eight years, shared the task of convening the British Sociological Association's Study Group on Sociology and Environment (incorporating Sociology and Architecture). During a time of restructuring upheavals in British higher education, this, with

xiii

the BSA's valuable support, provided a continuing and useful forum in which to focus attention on a neglected area of sociological research. And at the BSA office, Anne Dix deserves my very many thanks for her excellent support and assistance over many years.

The research library of the (late lamented) Greater London Council, the Educational Advisory Service of the Fulbight Commission in London, and the American Embassy in London also provided valuable information; I would also like to acknowledge a contribution from the Royal Institution of British Architects Research Fund in developing the initial framework for the research. Much of the library research was undertaken at the British Library (Lending Division) Boston Spa, and I would like to thank the Reading Room staff there for their excellent service.

Many other people have helped by sending papers or in other ways. I am very grateful to Barney Cohn, Harry Cowan, Frank Duffy, Michael Gilsenan, Shyam Gupta, Peter Hall, Nigel Harris, Deryck Holdsworth, Chris Husbands, Doreen Massey, Robin Murray, Saskia Sassen-Koob, Paul Rabinow, Ed Soja, Robert Thorne, John Walton, Goetz Wolff, and Ken Young. If Sydney King reads through the London case study he will see how much I have learnt from him concerning the international world of design.

At Routledge, I am indebted to Chris Rojek and John Urry (the series editor), for their help and valuable advice, and to Eve Daintith for her sympathetic and meticulous copy-editing.

Finally, to Ken Hall, whose careful word-processing as well as willing co-operation in burning the midnight oil have been instrumental in enabling me to complete this book on time, I would like to offer my sincere thanks.

Whilst I am happy to acknowledge the help of all the above, the final responsibility for the contents is, of course, my own.

Part One

CITIES AND THE WORLD-ECONOMY

Chapter One

INTRODUCTION
Cities and the world-economy[1]

GLOBAL PARADIGMS IN URBAN RESEARCH

The 1980s have seen a major paradigm shift in urban studies. The study of urbanization and the city has, like other phenomena, been directly linked to developments in the world-economy, the term 'global' becoming as common in book titles (on the global shift, global restructuring, global factory) as in the financial sections of newspapers. Theories of and studies on world or global cities (Cohen, 1981; Friedmann and Wolff, 1982; Friedmann, 1986), described by Feagin (1985) as 'the cotter pins holding the capitalist world-economy together' include work on Los Angeles, New York, Tokyo/Osaka/Ngoya, London, and São Paulo. The global context of metropolitan growth provides the framework for studies of, among others, Houston, Detroit, Buffalo, and, at a more popular level, Miami, Coral Gables, Paris, and Honolulu (Allman, 1983; Cooke, 1986b; Dickens, 1986; Feagin, 1985; Glickman, 1987; Grunwald and Flamm, 1985; Hall, 1984; Heenan, 1977; Henderson and Castells, 1987; Hill and Feagin, 1987; King, 1984ab; Kowarick and Campanario, 1986; Perry, 1987; Rimmer, 1986; Ross and Trachte, 1983; Sassen-Koob, 1984; 1985; 1986; 1987; Smith and Feagin, 1987; Soja, Morales, and Wolff, 1983; Soja *et al.*, 1985; Soja, 1986; Thrift, 1986b; 1987a; Trachte and Ross, 1985; Weaver and Richards, 1985).

In *Urbanization in the World-Economy* (Timberlake, 1985), the fifteen co-authors examine the book's theme in relation to different regions in the world. And from a more historical viewpoint, drawing on Wallerstein's (1987) world-system perspective, the effects of the historical incorporation of cities into the world-economy are discussed in relation to Latin America (Browning and Roberts, 1980; Meyer,

3

1986; Pang, 1983; Portes and Walton, 1981; Salinas, 1983; Slater, 1986), with a growing number of studies on Europe, Asia, and Africa (Armstrong and McGee, 1985; Chaichian, 1988; Chase-Dunn, 1985; Drakakis-Smith, 1986; 1987; Ewars, Goddard, and Matzerath, 1986; Henderson and Castells, 1987; McGee, 1986).

Other studies discuss the urban effects of the globalization of producer services (Daniels, 1986; Thrift, 1986ab); yet others (King, 1984a; Robertson and Lechner, 1985) the globalization of culture (from religion to aesthetics), the concept described by Robertson as 'the processes by which the world becomes a single place'. This refers both to the recognition of 'a very high degree of interdependence between spheres and locales of social activity across the entire globe' and, perhaps more importantly, 'the growth of consciousness pertaining to the globe as such' (Robertson, 1985: 348; see also Robertson, 1987; 1988; 1989).

Whilst these global perspectives on cities were already foreshadowed in the early 1970s (e.g. by Castells, 1977; Harvey, 1973; Wallerstein, 1974; Walton, 1976), this major shift in paradigms in the mid-1980s poses at least two questions. First, what changes are seen to have taken place in the objective, external world that account for this reorientation in urban research? Second, given that these objective changes have been taking place for at least three decades and, in the longer term, over three centuries or more, why has this theoretical and conceptual understanding of urban development gained currency only from the 1980s and not before? Why has the term 'global' assumed a new urgency? For example, in regard to the worldwide nature of economic activity, *The Economist* was advising its readers thirty years ago that 'today, more than ever before, exporters must plan *globally*, especially in view of the growing importance of the world's new markets' (5 October 1957); likewise, according to the Oxford English Dictionary (1984), the term 'globalization' had entered the vocabulary at the latest by 1962.

We shall deal with the second question first.

URBAN POLITICAL ECONOMY: AN HISTORICAL AND SPATIAL CRITIQUE

The 'new urban studies' (or urban political-economy approaches) revolutionized the understanding of and research on urbanization and urban development in the 1970s and 1980s. Drawing on the early work of Harvey and Castells, Walton clearly sets out some of the characteristics of this approach:

1. Urbanism and urbanization could not simply be taken for granted but required definition and explanation; they must assume the status of 'theoretical objects' in the sense that they arise (or do not) and take different forms under various modes of social and economic organization and political control.
2. The approach of the new urban studies is concerned with the interplay between relations of production, consumption, exchange, and the structure of power manifest in the state.
3. Actual or concrete urban processes, for example, ecological patterns, community organization, economic activities, class and ethnic politics (and one could add physical and spatial urban form, including architecture and urban design: author) must be understood in terms of their structural bases or how they are conditioned by the larger economic, political and socio-cultural milieu.
4. The approach is essentially connected with social change and sees this as growing out of conflicts among classes and status groups. Changes in the economy are socially and politically generated as well as mediated.
5. The perspective of the new urban studies is tied to the concerns of normative theory (of how things ought to be), concerned with both drawing out the ideological and distributional implications of alternative positions but also being critically aware of its own premisses.

(Walton, 1984: 78)

Yet with some notable exceptions (including Castells, Harvey, Roberts, Slater, Walton, and others) much of the new urban political economy (including urban social theory) has, as was argued in King (1989b), also been characterized by temporal, spatial, and conceptual restrictions; temporal, in the restricted historical dimension within which urban development is frequently analysed; spatial, in that despite recognition of (or in some cases, lip service paid to) the international context, national boundaries have too frequently been used to define the limits of a given urban system; and conceptual, in that relatively little attention has so far been given to the built environment, either as additional data for understanding social change and urban development, or, indeed, as a significant factor in influencing these phenomena. Whilst these comments may apply to the understanding of cities and urbanization in many countries, the following arguments draw particularly on urban studies in and of the UK.

The increased attention given to the global context of urban growth in the 1980s serves to emphasize its earlier neglect, a neglect not only of the historical but also of the geo-political context in which European and North American cities developed. One cause of this neglect has been the restricted focus of many studies in urban political economy on the problems of cities in the 'developed' economies of Europe and North America, as has recently been acknowledged (Harloe, 1987) and the (unstated) assumption that these could somehow be conceptualized separately from the rest of the world. The fact is that from the mid-1970s, the observable phenomena of a global system of production made their presence felt in two ways. The first was in the growing consciousness that with the expansion of multinational production (generally, though erroneously seen to have developed from the 1950s (see Dunning, 1983), jobs were being lost to countries where labour was cheaper (Blackaby, 1979; Singh, 1977; Portes and Walton, 1981). The second, and more visible factor was the high profile acquired by international labour migration ('immigrants', 'guest workers' (Castle and Kosack, 1978; Frobel, Heinrichs, and Kreye, 1980)), and the urban riots of 1980–1.

Among those concerned with urban development in the UK, these were some of the factors behind the increasing interest in 'the world outside', an interest that appears to have gathered momentum at this time: Massey (1986b), for example, dates this perspective from the Community Development Projects of 1977. Prior to this date, and even later, few studies on urbanization and urban development in Britain, whether in urban history, geography, sociology, or the 'new' urban political economy, paid much attention to this perspective as the journals and textbooks produced at this time will confirm (e.g. Cannadine and Reeder, 1982; Dyos and Wolff, 1973; Pahl, 1970; Pahl et al., 1983; Peach et al., 1978; Robson, 1973; Saunders, 1981). The work of Roberts (1978ab) is the major exception. Robson, for example, though recognizing the effects of international trade on urban growth in England and Wales, none the less undertakes his analysis on the assumption that these two countries form 'a closed system' (Robson, 1973: 46). 'Urbanization in developing countries' was treated as a distinct and separate phenomenon from 'British urbanization'

6

or, in cases, seen as a process that bore historical comparison to British experience, a view that has persisted into the 1980s (e.g. Johnson and Pooley, 1982). Whichever the viewpoint, the two were not seen as part of the *same* process. Thus, comments such as 'the periphery nations of Latin America, Asia, and Africa *now* become an accessible and seemingly inexhaustible reservoir of cheap labour' (Hill, 1984: 131, emphasis added) overlook the fact that millions of workers in those countries had, through indentured labour, the *encomienda* system, or simply as part of a colonial-plantation economy, formed part of the international division of labour for centuries. As in Wolf's study (1982), these are 'the forgotten people of Europe'. Smith and Feagin's recent suggestion (1987: 5) that urban development can best be understood 'by analysing cities in terms of their transnational linkages, especially their connections with the world capitalistic economy' is not only an agenda for the present (and future) but points up the inadequacy of our understanding of urban development in the past (and much urban policy analysis in the UK has been flawed for failing to recognize this (see King, 1986a; Thrift, 1985).

In short, one answer to the earlier question as to why there has been a sudden interest in globally oriented urban research in the 1980s is clear: only when the economic base of cities in 'advanced economies' at the core was affected, have many urbanists in those countries looked beyond national boundaries to the larger economic system that supported them. Yet the inhabitants of that 'external world' have long been aware that their own urban situation has been affected by core societies, except that this was seen as part of a larger process that went by another name — colonialism. Another answer is more prosaic: the academic realization that earlier explanations of urban growth and decline were too parochial and therefore, inadequate: the analysis required a deeper (historical) and broader (geographical) framework (see Harloe, 1987), which also went beyond traditional disciplinary boundaries.

Thus, whilst it is true that 'something fundamental happened' in the 1970s (Thrift, 1986b: 14) with major changes taking place in the organization of the world-economy (these are discussed in Chapter 2) equally significant has been a change in the *perception and understanding* of urban phenomena. Friedmann (1986: 69) attributes this to the 'special achievement' of Castells (1977) and Harvey (1973) in linking 'city forming processes to the larger historical movement of industrial capitalism'. Yet it is only since 1980, according to Friedmann, that 'the study of cities has been directly linked to the world economy'.

Whilst Friedmann's (1986) statement may be correct in terms of its precise wording (his use of the word 'directly') it is also misleading: fundamental to the development of the world-economy and the world system in general was the emergence of modern industrial colonialism, the cities that it created and through which it operated. Hence, the study of cities as 'directly linked to colonialism' is the necessary prerequisite for understanding the development of cities as 'directly linked to the world economy'. As indicated elsewhere (King, 1989b), most of the work cited by Friedmann in making his point is concerned with the impact of world economic processes on urban economies in the core states, especially in the USA. Earlier studies of colonial cities demonstrated the impact of industrial capitalism and world economic forces on city-forming processes (if not in the particular form or detail Friedmann (1986) implies) well before the date he suggests (e.g. McGee, 1967; Rayfield, 1974; see also King, 1989b).

However, equally important in the emergence of globally oriented urban research has been the selective adoption of the world-system paradigm (Wallerstein, 1974; 1979; 1984) and the erosion of disciplinary boundaries in urban research, particularly (and at the risk of caricature), between a more globally oriented 'development studies', previously focused on the periphery, and other academic disciplines previously focused on advanced industrial societies at the core (the division also relating to the work experience of practitioners in these fields); second, between social and spatial theorists (Gregory and Urry, 1985; *Society and Space*, 1983–). A third gap that has yet to be bridged is between social and spatial theorists and theorists of architecture and the built environment.

Other scholars, however (for example, Chase-Dunn, 1985: 273) do not see the recent changes noted by Cohen (1981) and Friedmann and Wolff (1982) as fundamentally new. Rather, trends that have been growing for the last 500 years such as the increases in the internationalization of capital, or in labour productivity, have continued and, due to rapid technological change, accelerated. The system of world cities did not simply appear in the 1950s with cities having been 'national' before that time: rather, cities have long performed both national and international functions.

This is close to the position adopted here, except that particular attention is devoted to the historic role of colonialism in accelerating the internationalization of capital and being instrumental in creating the present international city system.

Colonialism was (and is) a form of political economy that has spatial and built-form dimensions. In an earlier work (King, 1976) I attempted to show how this particular system of social and economic organization and political control gave rise to distinctive patterns of urbanization, urban development, and built environment in one particular colonized society. Though colonialism also affected the structure of urbanization and urban forms in the metropolitan society, this aspect was only hinted at in that account (1976, Chapter 2). It is developed further in *Urbanism, Colonialism, and the World-Economy* (King, 1989a) and also in this book because it is an essential prerequisite for understanding changes in the British urban system that are taking place today.

Since the middle of the twentieth century, the colonial mode of production — part of an old international division of labour in which British manufactured goods were exchanged for raw materials and food imports — has largely disappeared. The distinctive pattern of urbanization and urban and building forms to which it helped give rise have consequently undergone major changes. From the mid-twentieth century (and earlier) as the colonial empire disappeared, Britain has changed its role from being the major beneficiary in a colonial political economy to that of being a freer competitor in a larger, and harsher world-economy, a new international division of labour, albeit with particular connections to the European Community since 1973. What now needs to be explored is the way in which this new role has affected patterns of urbanization and urban and building form. Clearly such a task presupposes familiarity with a mass of data about the changing urban and regional system in relation to Britain's new role in the world-economy, data that has only recently been assembled (e.g. Cooke (ed.), 1986ab; 1989; Cooke and Thrift (eds), 1989). What this section attempts are some suggestions that may help undertake that project. More particularly, it has four aims:

1. To provide a more historically informed approach to the growing corpus of work on the global context of urban growth. Particularly, my interest is in exploring ways of understanding the recent development of building, architecture, and urban form on a world scale; however, the conceptual framework is of equal interest for examining social, economic, ethnic, racial, or cultural issues in general.

2. Linked to this is an attempt to understand the forces behind the production of building and urban form in world cities (the key centres in the world-economy) in general and social and physical forms in London in particular.

3. As these first two objectives require that attention be given to the historical phenomenon of colonialism, to examine the connection between an earlier colonialism and contemporary global capitalism, and between the colonial cities and world cities that each has produced. This also requires giving attention to theoretical formulations on both.

4. To pay particular attention to the role of the built environment, both as a product of and a contributor to the current phase of global restructuring. What role does the built environment play in promoting or inhibiting urban growth (or decline), encouraging new or reproducing old social relations and structures, or contributing to cultural continuity or change?

These four objectives clearly have an influence on the methodology adopted, not least the conceptualizations of 'urban systems' discussed in the next chapter.

Recent theoretical writing on world cities and their role and function in the world-economy (cited on p. 3) has considerable heuristic value. Yet whatever the particular functions of each world city, and the various structural characteristics it shares with others, the understanding of its distinctive economic, social, physical, and spatial characteristics requires a much deeper and broader analysis. Whilst Friedmann suggests that 'the economic variable is likely to be decisive for all attempts at explanation', he also expects cities to differ among themselves according 'not only to the mode of their integration with the global economy, but also their own historical past, national policies and cultural influences' (Friedmann, 1986: 69). It is these factors, and particularly, the phenomenon of colonialism as the mode in which London became incorporated into the global economy, that are explored here.

THE BUILT ENVIRONMENT

One of the more important gaps in knowledge concerns the role of the built environment (including architecture, building form, and urban design) in influencing urban and social change. Neglected in the new urban political economy, only recent debates on post-

10

modernism have directed attention to this sphere (Cooke, 1988a; Davis, 1985; Dear, 1986; Harvey, 1987; Jameson, 1984; Zukin, 1988).

Whilst space does not permit extensive discussion, it is clear from common-sense evidence that in recent years, the built environment has played a massive and central role in urban restructuring, even though the different aspects of that role have not been precisely examined. At the simplest level, economically, buildings provide for investment, store capital, create work, house activities, occupy land, provide opportunities for rent; socially, they support relationships, provide shelter, express social divisions, permit hierarchies, house institutions, enable the expression of status and authority, embody property relations; spatially, they establish place, define distance, enclose space, differentiate area; culturally, they store sentiment, symbolize meaning, express identity; politically, they symbolize power, represent authority, become an arena for conflict, or a political resource.

In the urban restructuring of the last decades, modifications to environments whether negative (through abandonment, neglect, or conscious destruction) or positive (in terms of new building, rehabilitation, and conservation) have been one of the most visible dimensions of economic and social change, inherently connected to sectoral changes in the economy, questions of employment, new working practices, and occupational structures.[2]

In core countries of the world-economy especially, but also in cities in general, urban restructuring has involved massive rebuilding and rehabilitation, the creation of vehicles for investment with 'value-added' provided by 'designer architecture' and the post-modern paradigm (Knox, 1987; Cooke, 1988b). In Britain, old 'country houses' are invested with new economic and social value (and meaning) by players on global financial markets (Thrift, 1987b). The new producer-services economy presupposes the existence of office towers just as the new competition in retailing presupposes consumption-oriented gallerias, shopping malls, and superstores. Likewise, in the developing countries of the periphery, self-built housing has become a profitable vehicle for international mortgage loans (Evers, 1984).

Whilst these preliminary comments make no attempt at systematic analysis, the built environment is given particular importance in the following chapters.

Chapter Two

WORLD-CITY FORMATION
An overview of recent research

INTRODUCTION

The growing body of literature on the 'world city' phenomenon has already been referred to. The purpose of this chapter is to make a brief review of this literature, to summarize the developments that are seen to account for the formation of world cities, and to provide a historical comment on these. Finally, the chapter looks at some of the main characteristics identified in the world city.

According to Hall (1984: 1) the idea of world cities as 'those in which quite a disproportionate part of the world's most important business is conducted' originated with Patrick Geddes in 1915. This conception, however, is something less than that implied in recent formulations. Braudel uses the term 'world-city' to denote the centre of specific 'world-economies' (Braudel, 1984: 26), an 'urban centre of gravity' as 'the logistic heart of its activity'. Whilst this concept is used in a historical context, it has more in common with recent conceptualizations.

Thus, Friedmann and Wolff are concerned with 'the spatial articulation of the emerging world system of production and markets through a global network of cities'. Specifically, their interest was in:

> the principal urban regions in this network in which most of the world's active capital comes to be concentrated, regions which play a vital part in the great capitalist undertaking to organize the world for the efficient extraction of surplus . . . the world economy is defined by a linked set of markets and production units, organized and controlled by transnational capital; world cities are the material manifestation of this control, occurring exclusively in core and semi-

peripheral regions where they serve as banking and financial centres, administrative headquarters, centers of ideological control and so forth.

(Friedmann and Wolff, 1982)

In such a system, there is, in the words of Sassen-Koob (1984: 140), the need for 'nodal points to coordinate and control this global economic activity'. The production of highly specialized services, top-level management and control functions constitute components in what she terms 'global control capability' (Sassen-Koob, 1986: 88) and the practice of global control is the specialized activity involved in producing and reproducing the organization and management of the global system of production and the global labour force (ibid).

For Ross and Trachte (1983), such cities are 'the location of the institutional heights of worldwide resource allocation', concentrating 'the production of cultural commodities that knit global capitalism into a web of symbolic hierarchy and interdependence'. In such places as New York City, London, Tokyo are:

the headquarters of the great banks and multinational corporations. From these headquarters radiate a web of electronic communications and air-travel corridors along which capital is deployed and redeployed, and through which the fundamental decisions about the structure of the world economy are sent. In these global cities work, but not necessarily reside, the cadre of officials and their staff who, in their persons and official capacities, embody the concentration and centralization of capital that now characterizes the global system.

(Ross and Trachte, 1983: 393–4)

More recent studies suggest that major cities tend to specialize in particular aspects of raw materials, production, distribution, marketing, financial, and other service activities (Smith and Feagin, 1987). Feagin's study of Houston indicates that, since the formation of OPEC in 1973, the city replaced its role of direct involvement in oil processing with a much greater emphasis on oil-related, often export-oriented, services technology, financial policy, and control functions. Such cities 'as the places where this politico-economic specialization is grounded physically, are the cotterpins holding the capitalist world economic system together' (Feagin, 1985: 30).

13

Cities such as these are seen to be at the apex of a new hierarchy of world cities that Cohen's definitive paper (1981) identified as emerging in the early 1970s. As multinational production and international finance became more dominant in Europe and Asia, an earlier hierarchy of national centres (see earlier editions of Hall, 1984) began to be displaced by a new hierarchy, characterized by corporate headquarters of multinational headquarters and banks.

THE RISE OF WORLD CITIES:
ALTERNATIVE ACCOUNTS

The massive increase in the internationalization of capital (or rather, of capitalist relations of production) over the last two decades is the common explanatory factor in all accounts of the world-city phenomenon (Cohen, 1981; Feagin and Smith, 1987; Friedmann and Wolff, 1982; Ross and Trachte, 1983; Sassen-Koob, 1984; 1986; Thrift, 1986b). The detailed study of restructuring in Los Angeles by Soja *et al.* (1983) provides the most comprehensive theoretical account of what they term the 'global capitalist city'. Both here and elsewhere, Soja (1986) draws on Mandel's (1978) arguments on long wave restructuring: at different historical periods, major and minor restructurings of capital have taken place, which have had different effects at different levels of the world economy (e.g. national, subnational, and city). In the last two decades, a much more dramatic internationalization of capital has taken place, leading to a more widely distributed system of production: the degree and extent of this change was most obviously manifest in the 1980 census figures and this prompted some of the earlier research (Soja, personal communication, 1987). The emergence of what Soja terms the 'global capitalist city', of the centre of financial management, international trade, and corporate HQ, has arisen from a series of economic and political crises marking the end of the post-war boom of the 1960s and the 'second slump' (1973–5) following the oil crisis. The crisis of overproduction has led to the intensification of the capitalist relations of production, with a deepening division of labour, the generation of new consumption needs, the incorporation of new spheres into capitalist production relations, and a greater concentration and centralization of capital. Using technological innovation, corporate managerial strategies, and state policies, restructuring has taken place to restore expanding profits and establish more effective control over the workforce; this has meant selective deindustrialization and reindustrialization linked to a strategy of anti-unionization (1983).

In discussing the new, world-economic order that has emerged since the 1970s, Thrift (1986b) focuses on the three 'principal groups of actors on the world stage' whose activities, collectively, have established the infrastructure for the emergence of the world city: multinational companies, banks, and state governments. Seeking international solutions to the national problems of the late 1960s, national firms have become international, and multinational firms, truly global (see below). In the 1980s, transnational corporations account for 70–80 per cent of world trade outside the centrally planned socialist countries (Feagin and Smith, 1987: 3). The extension and export of capitalist relations of production has taken three forms: multinationals have gained greater control over the world's raw materials; other countries' markets have been penetrated by producing in those countries, and cheap labour abroad has been exploited to produce goods for re-export to the home country of the multinational or to third markets. With the assistance of major innovations in telecommunications, data processing, and techniques of organizational management, this has led to the transformation of the multinational corporation (with subsidiary branches round the world) into the global corporation, with globally co-ordinated systems of production promoting global brand names. Such developments, at first mainly characteristic of American companies, have become more widespread from the 1970s; since then France, Australia, West Germany, the Netherlands, Japan, the United Kingdom, as well as an increasing number of developing countries have increased investments in manufacturing abroad.

The internationalization of production has been accompanied by the increased internationalization of finance, of which Thrift (1986b) lists three main components: the internationalization of domestic currency, with the massive growth in trading in exchange rates, and the growth of the Eurodollar market. The recycling of Middle Eastern oil surpluses, whether on a large scale, as loans to the 'better off' developing countries, with the subsequent social and political consequences such as food riots when loan terms were tightened (Walton, 1987) or, on a small scale, to fuel investment in time-share and leisure developments (King, 1984a; Gulf Leisure Investments, 1986) are both indications of the local outcome of global investment practice. Second, aided by spectacular developments in communications, banking has been internationalized, becoming 'truly global' in the 1970s, as also have capital markets, with 24-hour global trading in securities; stock exchanges, futures exchanges, and commodity markets have proliferated all round the world.

There has also been, as Thrift (1986b) (from whom much of this account is drawn) points out, an increased internationalization of the state, which in many cases, has gone out of its way to attract foreign investment and foreign companies, policies promoted by international organizations such as the IMF, the World Bank, and the OECD. The cumulative effect of these developments has been that capital has become more footloose; there has been a greater interpenetration of capital so that the world-economy is not now simply based on a single economic pole (the United States) but has become multipolar, with the borders of capitalist production moved much further out, and encompassing an increasing number of newly industrializing countries. It is on the growth of these (especially Taiwan, Hong Kong, Singapore, South Korea, Brazil, Mexico) that Harris predicts *The End of the Third World* (1987).

This phrase is perhaps more aptly applied to developments in the world labour market in the 1980s where increasingly links between First- and Third-World workers are forged, not least in world cities. The new international division of labour (Frobel, Heinrichs, and Kreye, 1980), which has accompanied, and made possible the internationalization of production, has seen the massive growth of people directly employed by multinationals outside their countries of origin: between 1960 and 1980, the proportion of people employed by American multinationals outside the USA rose from 8.7 to 17.5 per cent (see also Portes and Walton, 1981), though the sectors of employment were mainly in textiles and electronics. These were developments helped by the increased fragmentation and spatial separation of the production process in order to take advantage of areas of unskilled and low-paid labour. Along with Export Processing Zones (the first established at Shannon International Airport in 1956), which bring work to the workers, the massive growth of international labour migration in recent years has brought workers to the work. In 1979, the USA had some 5 million legal and between 2.5 and 4 million illegal migrants. In 1979, Western Europe had some 6.3 million, and Arab countries almost 3 million migrant workers; the USA had some 5 million migrants. But developing countries have also seen an increase in the same phenomenon, with 3.5 to 4 million migrants in Latin America and almost 1.5 million in West Africa (see also Cohen, 1987).

16

All these developments have been associated with, and dependent on, the growth of the international service economy, based on corporate activities. Internationalization of production and finance has meant the internationalization of administration and control through advanced producer services, activities assisting user firms to carry out administrative, development, and financial functions, whether these are research and development, strategic planning, banking, insurance, real estate, accounting, legal services, consulting, advertising, and so forth. It is this that has extensively changed the employment structure in the 'advanced' capitalist countries and it is the growth of such activities and employment that Thrift (1986b) sees as being intrinsic to the formation of world cities (p. 60).

Fundamental shifts have taken place in the conditions of the 'new competition' of advanced capitalism. In the poor countries on the global periphery, cities have gained importance as industrial centres involved in volume production for export to markets in core countries. In the rich nations, according to Noyelle (1986), as everyone has more than enough to live on (despite its maldistribution), the emphasis has moved away from production to 'product development', distribution, marketing, selling, and — not least — advertising. 'Hype', 'value-added', and 'designer' have become the characteristic terms of the 1980s. Companies have exhausted many of the possibilities for economies in the production of goods and therefore increasingly look for economies in administration and service functions. In Gutman's view, it is this that has benefitted architectural services as these have been devoted to lowering the cost of construction and maintaining facilities, plants, and buildings that firms and organizations require for their operations (Gutman, 1988). 'Facilities management' (Duffy, 1983) has, therefore, become yet another producer-service subspecialization.

Hundreds of mergers in recent years have accompanied these developments, cutting costs for global competition, increasing the extent of monopoly yet again, whether for merchant bankers, advertising agents, or the designers of glossy takeover brochures, bringing increased business to particular service businesses.

It is not only the growth but also the *export* of these producer services that has led to world-city growth (Sassen-Koob, 1984). As the comparative advantage of Western advanced industrial countries in the production of goods has declined, they have turned to alternative sources of profit. This has led them to exploit the possibilities offered

17

by service industries, especially those (like banking, insurance, design services, accounting, etc.) where technological superiority in information and data processing gives the comparative advantage back to them. It is this that has led governments in such countries to press for liberalization in the international trade in services. It is this, too, that keeps information in the world cities at the core. The post-industrial trade war is fought over information.

Thrift (1986b) divides such world cities into three categories: first, the truly international centres (New York, London, Paris, and Zurich) containing many head offices, branch offices, and regional headquarters of large corporations and representative offices of many banks. Second are the zonal centres (Singapore, Hong Kong, and Los Angeles) serving as important links in the international financial system but responsible for particular geographic zones rather than world-scale business. Finally, the regional centres (Sydney, Dallas, Chicago, Miami, and San Francisco) host to corporate headquarters and foreign financial outlets but not essential links in the international financial system (p. 61).

WORLD CITIES: AN HISTORICAL PERSPECTIVE

What is missing from these accounts is a historical perspective. Such world cities have, as Braudel indicates (1984), not simply emerged since the 1960s nor, despite the spectacular growth of communications and transport in the last two decades, is the concept (as opposed to the phrase) of 'global control capability', one that has only recently surfaced. Much more needs to be known about the earlier infrastructure on which contemporary systems of economic, political, and especially, cultural control and hegemony have developed.

For example, work on the historical background of multinationals suggests that by 1914, at least 14 billion dollars had been invested in enterprises or branch plants where non-resident investors owned a majority or large part of equity interest, or that were owned or controlled by first-generation expatriates who had earlier migrated (Dunning, 1983). This amount represented about 35 per cent of the estimated total long-term international debt at that time. 'There is little doubt', writes Dunning (1983), 'that several economies, particularly those of developing countries, and especially capital-intensive primary producers and technology-intensive manufacturing sectors, were dominated either by affiliates of MNEs or by foreign enterprises.'

Whilst the first seventy years of the nineteenth century saw mainly direct capital investment, the forty years between the 1870s and 1914 saw 'the infancy and adolescence of the type of activity which mainly dominates today, that is, the setting up of foreign branches by enterprises already operating in their home countries' (Dunning, 1983: 86).

In these activities, the United Kingdom, with over 45 per cent of the accumulated foreign direct investment in 1914, was by far the largest capital stake holder, with the USA (about 18 per cent) far behind. Four-fifths of all this investment was in what today would be described as developing countries, outside Europe and the USA. A good half of this investment was in the primary-product sector, 20 per cent of it in railroads, and 15 per cent in manufacturing activities. Apart from iron ore, coal, and bauxite, almost all mineral investments were located in the British Empire or in developing countries.

As Christopher shows (1988: 67–81), with the *Companies Act* of 1858–62 enabling limited liability companies to be formed, there was a vast outflow of investment capital to the rest of the world. Between 1870 and 1913, British investors placed 5.27 per cent of total GNP in overseas lending (Edelstein, 1981); 'no other country approached this level of external lending and it has not been repeated since'. Almost 70 per cent of all this new issue capital was invested in public utilities: railway systems had 41 per cent of the total, with a further 28 per cent on tramways, docks, telegraphs, telephones, gas, electricity, and water works. This was the infrastructure of what was later to become, in many cases, the 'global city' of the semi-periphery.

Of the £3,763 million invested overseas in 1914, almost half (£1,780 million) was invested in the Empire, the rest in foreign countries elsewhere. The main recipients are shown in Table 2.1. Whilst the banks in the colonies were at first locally controlled they were later (1860s to the 1870s) taken over by banks in London (Christopher, 1988: 70). (See also Chapter 5, p. 89.)

Particularly significant at this time were the raw-material and agricultural investments: 'this was the heyday of the plantations, e.g. rubber, tea, coffee and cocoa; of cattle raising and meat processing, e.g. in the USA and Argentine; and of the emergence of the vertically integrated MNE in tropical fruits, sugar and tobacco' (Dunning, 1983: 88). Apart from some transnational railroad activity in Europe and Latin America, it was mainly in the agricultural sector that this international hierarchical organization made itself felt, particularly in economies where prosperity rested on a single crop, where the production and marketing was controlled by a few (and sometimes only one) foreign companies, for example, Cuba (sugar), Costa Rica (bananas), Ceylon (tea), and Liberia (rubber).

Table 2.1 British overseas investment, 1914

Empire	£million	Other foreign countries	£million
Canada &			
Newfoundland	515	USA	755
India & Ceylon	378	Argentine	319
South Africa	370	Brazil	148
Australia	332	Russia	110
New Zealand	84	Mexico	99
West Africa	61	Chile	61
Malaya	45	Egypt	45
British N. Borneo	6	China	44
Hong Kong	3		
Other colonies	26	Other countries	402

Source: Christopher, 1988: 68.
Note: Figures to nearest million.

As Dunning points out, there were distinct geographical and industrial patterns of foreign direct investment that varied with the home country of the investor. 'Language, cultural, political and trading ties, as well as geographical distance played a more important role than they do today' (Dunning, 1983: 90).

Further foreign direct investment developed in the years between the wars, as the share of US multinationals of the world capital stock rose from over 18 per cent in 1914 to almost 28 per cent in 1938. New multinationals invested in oil in the Mexican Gulf, the Dutch East Indies, and the Middle East; copper and iron ore in Africa; and bauxite in Dutch and British Guyana. The increased demand for rubber saw American and European multinationals invest in plantations in Liberia, Malaysia, and the Dutch East Indies; and rising living standards at home prompted further investment in sugar, tropical fruit, and tobacco. During this period, investment by Continental European firms was mainly in Europe while US firms were strongly oriented to Latin America, Canada, and the larger European countries. The first Japanese manufacturing affiliates were set up between 1920 and 1938. These data are sufficient to demonstrate a sophisticated programme of transnationalized production long before 1950 (Dunning, 1983: 85–93).

Similarly, a world-city-based system of international banking was instrumental in incorporating less-developed countries into the world economy in the nineteenth and twentieth centuries (Yannopolis, 1983). Like other Latin American countries, the Argentine provides a good example. In the 1860s and 1870s, British mercantile houses were heavily involved in land investment, railroad building, and other services in Buenos Aires; by 1900, four British houses exported 75 per cent of Argentine grain (the railroad system had been particularly oriented to export functions). Buenos Aires was transformed into a specialized bureaucratic and commercial metropolis impeding the subsequent industrialization of the Argentine (Pang, 1983). In Brazil, Rio de Janeiro followed a similar pattern (Ribeiro, 1989). As Roberts has pointed out (1978a), the commercialization of staples like coffee or sugar in Brazil and Cuba, tended to create urban centres whose growth was conditioned by the needs and demands of the countries at the core. These cities provided the sites for the international banking system headquartered in London, New York, and European capitals.

The Bank of London and South America (founded 1862), the British Bank of South America (1863), and the Anglo-South American Bank (1888) were all based in London. In the early 1930s, the Anglo-South American Bank had over forty branches throughout Latin America (including eight sub-branches in Buenos Aires) (as well as other branches in Bradford and Manchester); the Bank of London and South America had its main office (and eight suboffices) in Buenos Aires and branches at another dozen towns in Argentina, seventeen in Brazil and others throughout the continent. Seven major American banks, headquartered in New York, operated on the continent (the Bank of America, the National Association, the Chase National Bank, the Chemical Bank, the First National Bank of Boston, the Guaranty Trust Company, the Irving Trust Company, the National City Bank of New York): the latter, with some two billion dollars total resources, had branches all over the continent, including twenty-six in Cuba alone (the Royal Canadian Bank had about the same number there). Other banks headquartered in world cities included four from Montreal, two from Paris, two from Berlin as well as other international banks with headquarters in Antwerp, Amsterdam, Milan, Taipeh, Yokohama, Lisbon, and Basle (Howell Davis, *The South American Handbook*, 1934).

Apart from the local and expatriate representatives of these banking firms, other forerunners of today's producer-services community

21

included estate agents, insurance firms, shipping and air agents, advertising firms, real-estate firms, managing agents, and others.

The immediate source of this data, the official handbooks and guide books to various regions of the formal and informal colonial world, provide, in their advertisements and contents, invaluable clues to the uniquely specialized, complementary, and essentially interdependent nature of the economic, social, industrial, urban, or building developments in particular localities — whether in the metropole or the colony — which are evidence of the particular international division of labour at that time. Thus, a guide to the West Indies from 1914, when sugar-cane production dominated the islands' economy, carries advertisements of five major manufacturers of sugar-cane processing machinery, four of them in Glasgow; the head offices of Jenkinson, Brinsley, and Jenkinson, chartered surveyors and auctioneers, 'land agents for English, Colonial and West Indian estates, established for over 100 years' (i.e. before 1814) are in Ludgate Circus, London EC1; advertisements for five colleges or schools catering for colonial officials resident in the West Indies indicate locations at Petersfield (Hampshire) ('for children of colonial and British parents'), Folkestone (Kent), Southampton ('colonial children receive special care and attention'), and Southport (Lancashire). At Upper Norwood in South East London, a 'finishing school for daughters of gentlemen' indicates that such appendages of the colonial system are both spatial and social: here, a staff of qualified resident English and foreign governesses and visiting professors prepared 'young ladies to take their place in Society under less irksome conditions than ordinary school discipline', the school taking 'entire charge of pupils whose parents are abroad' for a fee from £135 a year.

The Colonial Bank (incorporated 1836) has its head office in Bishopsgate, London, and branches in all the West Indian islands. Street and Co., advertising agents, of 30 Cornhill, London 'has been established over 70 years (i.e. in the 1840s) and has among its clients some of the most substantial houses in the British Isles' (Aspinall, 1914). Elsewhere they describe themselves as 'the leading consultants for advertising in the Empire and throughout the world, with offices in New York, Paris, and Johannesburg' (*The South and East African Year Book and Guide*, 1936; G.G. Brown ed., 1938). The Bahamas, 'The ideal resort, a land of roses and sunshine' with 'sea bathing of unrivalled excellence, boating, fishing, golf, tennis, cricket, cycling, driving, are only ten days from England, via New York'. Elsewhere shipping companies and agents, tourist organizations, air services, periodical and book publishers, specialist tropical clothing suppliers, and food and drink exporters, hotel, restaurant, and club proprietors, map makers, automobile manufacturers, railways, shipbuilders, all constitute major advertisers whose living largely depend on the overseas and colonial connection, and whose potential fortunes depend upon, to a very large extent, the relevant specialized activity on the colonial periphery. (For the Great Western Railways, the South Wales Ports (Newport, Cardiff, Penarth), are promoted as 'the shortest route between Birmingham and the Midlands and Overseas', *The South American Handbook*, 1934).

Thus, just as in Pang's (1983) 'bureaucratic and commercial metropolis' of early twentieth-century Buenos Aires there were four-teen English schools, eleven Freemasons' lodges, ten Protestant churches, a British hospital, the Boat Club, the Jockey Club, the English Literary Society, the Hurlingham Club, the Victorian Tea-Rooms, eight English newspapers, about fifteen hotels (including the Palace, the Grand, the Royal, the Chester, the Albion, the Garden, the London, the Plaza, the Phoenix, the Metropole, the Splendid, the Provence, and the Brunswick and the Sportsman restaurants), as well as specific suburbs (such as the Belgrano with 'a very large propor-tion' of English and German residents, the Belgrano Athletic Club and a literary and debating society, or the Banfield, 'very popular with English-speaking people', with its lawn tennis courts (*Mitchell's Guide to Buenos Aires*, 1910) and no doubt domestic architecture), to provide for the British antecedents of the 50,000 Anglo-Argentinians who were registered at the time of the Anglo-Argentine conflict in 1984, and who, in 1910, managed, monitored, and partly controlled the economic, political, and cultural connection to Britain, so, in the British metropole itself, equally specialized institutions, organizations — and environments — existed, dependent on the Anglo-Latin American con-nection. Similarly, there were others dependent on the link with Anglo-Indians, Anglo-Africans, Anglo-Americans, or Canadians. The one would not exist without the other.

It is hardly necessary to mention that such institutions, though now generally transformed, and whether in Britain or in other once-colonial or subcolonial countries, still persist, though now maintained by different economic, social, or political supports. Moreover, these are not just institutions in the form of organizations, offices, roles, or prac-tices (such as schools, colleges, government departments) but also historically and socially produced attitudes, beliefs, values, networks, distinctive forms of knowledge, having influence at both the national and international level, and being expressed both spatially and physically as one layer of investment in the economic geology of the country (Massey, 1984). In Britain, the most obvious examples are the private boarding schools and colleges, now sustained by financial injections from new international elites, naval and military academies, merchant banks, and commercial firms, ideological or religious organizations founded on earlier politico-missionary activity, and professional and social institutions still maintaining networks round the world.

Such institutions, networks, and organizations, to a greater or lesser extent transformed, persist in the metropolitan centre, concentrated in particular regions, sustaining particular populations and environments. They are part of a past imperial, but also present international geography, elements in not only a spatial and social but also a cultural international division of labour. They also provide the infrastructure for new institutions, elements in the new international division of labour. There is change — but there are also continuities.

WORLD-CITY CHARACTERISTICS

It is their enhanced function as 'command and control centres' in the capitalist world-economy that, on the one hand, provide the selection criteria for world cities and, on the other, give rise to the characteristics attributed to them. In setting out these characteristics, the first purpose of this section is to provide a background against which to look at recent developments in London and its role as world city (See Chapter 5); second, to compare the concept and characteristics of the world city with those of the historic colonial city both for theoretical as well as more practical purposes.

The following discussion draws from the general accounts of world-city formation by Cohen (1981), Friedmann (1986), Friedmann and Wolff (1982), and Feagin and Smith (1987) as well as studies of individual cities, particularly New York, Los Angeles, Tokyo, and to a lesser extent, São Paulo and London (Light, 1988; Rimmer, 1986; Sassen-Koob, 1984; 1985; 1986; 1987; Sternlieb and Hughes, 1988; Soja et al., 1983; Soja et al., 1985; Soja, 1986; Thrift, 1986a). It is, therefore, particularly weighted by data from American cities and in these accounts the role of the state in promoting or inhibiting these recent structural changes has received less attention.

24

Selection criteria

Friedmann (1986) uses seven major criteria to identify thirty cities that he subsequently arranges in the 'world-city hierarchy'. These include primary and secondary cities in core countries (e.g. London and Brussels) as well as primary and secondary cities in semi-peripheral countries (e.g. Singapore and Johannesburg). Core countries are identified by World Bank criteria as industrial market economies and semi-peripheral ones, as upper-middle-income market economies (see Table 2.2).

Table 2.2 The world-city hierarchy

Core countries			Semi-periphery countries		
Primary		*Secondary*	*Primary*		*Secondary*
London†	***	Brussels† *			
Paris†	**	Milan *			
Rotterdam	*	Vienna† *			
Frankfurt	*	Madrid† *			
Zurich	*				Johannesburg *
New York	***	Toronto *	São Paulo	*	Buenos Aires† ***
Chicago	**	Miami *			Rio de Janeiro ***
Los Angeles	***	Houston *			Caracas† *
		San Francisco *			Mexico City† ***
Tokyo†	***	Sydney *	Singapore†	*	Hong Kong **
					Taipei† *
					Manila† **
					Bangkok† **
					Seoul† **

Source: Friedmann (1986) 'The World City Hypothesis', *Development and Change* 17 (1): 72.
Note: † National capital.
Population size categories (recent estimates referring to metro-region)
* 1–5 million   ** 5–10 million   *** 10–20 million

The criteria Friedmann (1986) uses to indicate world-city status are: major financial centre; site for headquarters for transnational corporations (TNCs), including regional headquarters; international institutions; rapid growth of business-services sector; important manufacturing centre; major transportation node; population size. (Other world-city analyses, for example, Rimmer, 1986 and Thrift, 1985, include other criteria such as research and education centre, convention and exposition function, etc.)

The most inherent feature of the world city is its global-control function and this gives it its principal geopolitical characteristic.

1. 'World cities lie at the junction between the world economy and the territorial national state' (Friedmann and Wolff, 1982).

The institutions linking the world-economy to the state are the transnational corporations and international finance, hence:

2. Major cities of core countries contain a disproportionate number of the headquarters facilities of the world's 500 largest national corporations as well as the offices of international banks.

In 1984, the four world cities with over twenty TNC headquarters each were New York (59), London (37), Tokyo (34), and Paris (26) (though Rimmer (1986) quotes data indicating Tokyo with a higher number than New York). Six other cities all had over ten headquarters: Chicago (18), Essen (18), Osaka (15), Los Angeles (14), Houston (11), Hamburg (10), Dallas (10) (Feagin and Smith, 1987).

Linking these corporate headquarters to international markets and articulating the global financial system are the major banks. Between 1970 and 1984, the number of foreign banks directly represented in New York rose from 75 to 307 and in London, from 163 to 403 (*The Banker*, November, 1984).

The presence of corporate headquarters, global financial and international institutions presupposes:

3. The rapid growth of a highly-paid international élite, including a transnational producer service class (in law, banking, insurance, business services, accounting, engineering, advertising, etc.) engaged in the production and export of services from world cities. In 1981, producer services accounted for 31 per cent of all employment in New York City and 25 per cent in Los Angeles (Sassen-Koob, 1984). Between 1970 and 1980, a rapid growth in international services trade saw service exports increase at an average rate of 19 per cent a year compared to 20 per cent in goods: in 1980, the top two American advertising firms made over 50 per cent of their income from abroad (as also did the top ten accounting firms) (ibid: 145). Between 1977 and 1980, the Los Angeles region saw a 20–30 per cent increase in major service industries, including management, public relations, engineering and architecture, and protective services). In roughly the same period, engineering services grew by over 60 per cent in New York, 25 per cent in Los Angeles, and computing and data processing by over 60 per cent in New York and 41 per cent in Los Angeles. Educational and medical services are also expanding.

4. 'As centres of global transport, communications, the production and transmission of news, information and culture, world cities perform important ideological and control functions' (Friedmann, 1986).

5. The expansion of global management and financial services functions, including growth of producer services, has led to the refurbishment of existing office space and extensive new construction both of office space and high class residential accommodation. Between 1980 and 1981, the total level of construction in New York City went up by 7 per cent compared to 1 per cent nationally; prior to 1960, downtown Los Angeles contained few office buildings, but between 1960 and 1970, the city moved from ninth to fifth in the US on the list of corporate headquarter cities. Over 30 million sq ft of office space was constructed between 1972 and 1982, a 50 per cent increase in ten years (Soja *et al.*, 1983; Light, 1988).

6. Because of the expanding economy, redevelopment, and concentration of capital, world cities become major centres for international investment.

In 1980, about one-third of the most valuable properties in the western edge of downtown Los Angeles were owned by foreign companies. By 1988, 70 per cent of downtown Los Angeles was foreign-owned (Wald, 1988: 8).

7. There has been a decentralization of routine office jobs and recentralization of control and management functions in the world city.

8. Increased global competition and falling rates of profit have led to the rationalization of manufacturing bringing de-industrialization in unionized, blue-collar employment. To a much lesser degree, there has been selective re-industrialization in high tech, lower wage, assembly line jobs, not unionized and often held by immigrant labour or ethnic minority women.

Between 1970 and 1980, New York City experienced an absolute decline in employment from 3.7 to 3.3. million, with a 35 per cent loss in manufacturing jobs; in the same period, the Los Angeles region gained some 226,000 manufacturing jobs as its total population grew by 1,300,000.

9. The growth of this high level service sector leads to expansion of some white collar employment but has had much greater multiplier effects for the growth of low-level, low paid service jobs, needing little skill or language proficiency and with few possibilities

of advance, increasingly undertaken by non-unionized, often immigrant and female labour.

These jobs are in the fastest growing employment sectors (maintenance, office cleaning, tourism, clerical, sales employment, hotels, restaurants, domestic service), the dynamic sector of the economy.

In the US in general, between 1960 and 1975, whilst there was a 35 per cent increase in jobs in the two highest earning classes and only an 11 per cent increase in the medium one, a 54 per cent increase took place in the two lowest earning classes (Sassen-Koob, 1984; 1986).

10. Much of the growth in industrial employment is associated with informally organized types of manufacturing such as sweatshops, industrial homework, both in traditional (clothing/garment) as well as new (electronics) industries, again undertaken by immigrant labour and ethnic minority women.

11. Depending to various degrees on the application of state controls, there has been a new influx of immigration directed to a few major urban centres which are seen as providing employment opportunities and this immigration has been accompanied by a growth in low living standards.

In the early 1980s, half of all the immigrants lived in the 10 largest American cities compared to only 11 per cent of the US population as a whole. In the last twenty years, 2 million Third-World migrants from South-East Asia, the Middle East, and Latin America have moved into the Los Angeles region resulting in a major change in the ethnic, economic, and social composition of the city (in 1953, Los Angeles County was 85 per cent Anglo-Saxon; in 1983, Hispanics, Blacks, and Asians comprised 50 per cent of the population). In the early 1980s, 39 per cent of Hispanics in New York City were below the official poverty line; 50 per cent of these were labourers, service workers, and operatives. Ross and Trachte (1983) quote immigrant wards in New York City where birth rates and infant-mortality figures are comparable to Third-World levels. Soja et al. (1983) compare parts of Los Angeles to a 'Third World metropolis'; Sassen-Koob (1984), referring to the same aspects, speaks of 'the peripheralization of the core'.

12. The outcome of these structural employments trends, combined with the effects of immigration has been a sharp increase in income, occupational and social polarization.

13. This polarization of high pay/high skill and low pay/low skill

28

employment is reflected in an increasing spatial separation of residential space according to occupation, race, ethnicity and income. The degree of ghettoization has greatly increased.

14. The growth of a high income class provides the conditions both for new luxury residential building and also gentrification of older areas (Williams and Smith, 1985).

In Los Angeles, the Bunker Hill redevelopment scheme replaced over 7,000 low-income residences and many small businesses, making way for an opera house, corporate headquarters, and luxury housing (Light, 1988). In London, the redevelopment of Docklands, a traditional working-class area, by luxury housing follows the same pattern (See Chapter 5).

15. Likewise, the combinations of conditions outlined above (national and international immigration, economic and social polarization, unemployment) as well as redevelopment and rapidly increasing competition for buildings and land causes major crises in housing.

In the early 1980s, there were some 30,000 homeless in Los Angeles and equally serious problems of homelessness in New York. By 1988, New York City's homeless population was 28,000 (*New York Times*, 15 May, 1988, p. 30).

16. The growth in low skill, low pay employment combines with a surplus of labour in the city to inflate the informal, floating or street economy. To these are added the numbers of unemployed, dependent on family support and public charity. Increasingly, an underclass exists of different ethnic origin from the ruling strata, giving many world cities a 'Third World' aspect.

17. There is a growing loss of local control over, and regulation of an increasingly footloose capital, and this is combined with increasing public expenditure to attract or maintain capital investment.

18. Major growth sectors in the urban economy are accompanied by equally major declines in others; new expansion is paralleled by accelerating decay.

19. Heightened political conflict results from contradictory demands of forces representing international capital and the needs of the local state. But conflict also arises over the restructuring of employment, housing, racial and ethnic tensions.

20. Increasing economic and social polarization, floating population and other conditions combine to lead to increased lawlessness

expressed in racial conflict, violence, increased crime rates, and street demonstrations, despite increased expenditure on police and a growth in private security forces and expenditure.

It is these characteristics, initially explored by Friedmann and Wolff (1982) that have been revised and reformulated by Friedmann (1986) as a series of hypotheses, the aim of which are 'to link urbanization processes to global economic forces'. Primarily intended as a framework for research, the theses are meant as a starting point for further enquiry, as an examination of individual cities would obviously expect to find significant differences among those which have become 'basing points' for global capital. Also, cities would clearly differ not only according to their mode of integration within the global economy but also depending on their historic past, national policies, and cultural influences. None the less, as already indicated, Friedmann concludes 'the economic variable is likely to be decisive for all attempts at explanation' (1986: p. 69.).

Though a proper understanding of the seven interrelated theses obviously requires the commentary that Friedmann attaches to them, for reference purposes they are stated here without further elaboration:

1. The form and extent of a city's integration with the world economy and the functions assigned to the city in the new spatial division of labour, will be decisive for any structural changes occurring within it.

2. Key cities throughout the world are used by global capital as 'basing points' in the spatial organization and articulation of production and markets. The resulting linkages make it possible to arrange world cities into a complex spatial hierarchy.

3. The global control functions of world cities are directly reflected in the structure and dynamics of their production sectors and employment.

4. World cities are major sites for the concentration and accumulation of international capital.

5. World cities are points of destination for large numbers of domestic and/or international migrants.

6. World city formation brings into focus the major contradictions of industrial capitalism — among them, spatial and class polarization.

7. World city growth generates social costs at rates that tend to

exceed the fiscal capacity of the state.

(Friedman, 1986)

Friedmann's hypothesis has subsequently been the subject of a critique by Korff (1987). Some of the points made by him may be briefly mentioned, together with an equally brief response.

1. If world system analysis is to be taken seriously, the mechanism and influences of the world system have to be analysed down to the smallest units. The units not only consist of 'world cities' but also 'world villages' and 'world households'.
This is a valid point and substantially coincides with issues raised and with the methodology, drawn from earlier work, proposed in Chapter 4.
2. The starting point for the world city hypothesis should focus on the detailed analysis of specific cities rather than at the macro level of world system analysis.
This misunderstands the function of a hypothesis, the main purpose of which is to put it to test against empirical reality.
3. The city's development is not only the result of world system processes but of regional and inter-regional patterns of trade which tend to be neglected in world city analysis.
Whilst this might seem a valid issue, these dimensions are not necessarily excluded from world-system analysis.
4. A more valid criticism is that in the modern world economy (particularly in low-income countries, for example, those mentioned by Korff), not all sectors can be classed as belonging to a system of capitalism.
5. In Europe or Japan whole countries or areas in their entirety act as 'world cities' and are centres of the world system (this point has also been hinted at by Rimmer, 1986).
This is also a valid point. Though Friedman (1986) uses the concept of 'metropolitan region', in many European countries, the integrated nature of urban and communication systems means that (as in the UK), the 'world city' extends to large parts of the country.

Whilst other comments made by Korff provide useful points of entry into the investigation of political and social processes in world cities and also emphasize the critical importance of making a detailed historical examination of given countries and cities, none of these inputs was excluded from Friedmann's original hypothesis. It remains a useful 'framework for research (and) . . . a starting point for political enquiry' (Friedmann, 1986: 69).

The more critical issue is in his statement that 'the economic variable . . . is likely to be decisive for all attempts at explanation'. Clearly, if one starts with the conceptual assumption about a 'world economy', such a statement will be true. If the starting point was a 'world culture' other variables would be decisive (see 'Introduction', King, 1989b).

Chapter Three

WORLD CITIES, COLONIAL CITIES
Connections and comparisons

WORLD CITIES, COLONIAL CITIES: CONNECTIONS

What is the connection between contemporary world cities and historic colonial cities and what comparisons can be made between them? The purpose of asking these two questions is to throw some light on the answers to three others: first, if we accept the criteria used for their selection, why have some cities attained the status of world city and not others? Second, what differences are there, in terms of functions or structure, between different world cities in the contemporary capitalist world-economy? Third, what factors are likely to be important in determining which cities take on world-city status in the future?

In the following comments — which attempt to explore rather than furnish answers to these questions — the term 'colonial city' is used in a more comprehensive sense than previously (King, 1989b). Taking the empire as a single political and economic unit, it is taken to refer to one of three categories: the metropolitan capital of a colonial empire (conventionally speaking, an 'imperial city' e.g. London, Amsterdam, and Paris); the capital city of a colony within an empire (Delhi, Elizabethville, and Rabat); colonial port cities and regional administrative centres. In each case, the size, functions (economic, social, administrative, political), social, cultural, and spatial form can be understood only by reference to its specific role within the larger colonial political economy as well as the larger world system.

While opinions may differ concerning the selection of cities included in Friedmann's formulation of the world-city hierarchy (Table 2.2, p. 25) (for example, Rimmer, 1986; Thrift, 1986a: 61), the list can nevertheless be taken as illustrative. In which case, the first point to notice is that about half of the cities cited (and particularly those in the semi-periphery) were previously either metropolitan capitals of European colonial empires (e.g. London, Paris, Brussels, and Madrid — though Amsterdam is not included) or either zonal or regional centres within these empires: Sydney, Toronto, New York, Singapore, Hong Kong, Buenos Aires, Caracas, Mexico City, Manila, and Jakarta. Hence, this already suggests that the status of a city in the contemporary world hierarchy has much to do with its previous status in a colonial hierarchy and colonial system of production.

The reasons for this are obviously historical and include a wide range of factors — economic, political, military, administrative, religious, and cultural, which have been discussed in *Urbanism, Colonialism, and the World-Economy* (King 1989b) (see especially Chapter 1). The over-riding factors holding the capitalist world-economy together are shared ideology, economic and political power, and communications, in terms of both telecommunications and air transport, with the network routes of the latter resulting more from the geopolitics of history (and reflecting past colonial relationships) than from spatial or commercial logic (Jenkins, 1971).

As set out in *Urbanism, Colonialism, and the World-Economy* (King, 1989b), therefore, colonial urbanization and cities became the instruments by which the colonial periphery was incorporated into the metropolitan core. For each of the main colonial powers, a colonial urban system was established, from the metropolitan capital and port cities, through a network of colonial port cities, colonial capitals, regional and district centres, down to the outlying stations of the colonial bureaucracy and system of military control. All were linked by transport, communications, and subsequently, electronic and other media. As pre-capitalist modes of production, often with little urbanization or towns, were replaced, new capitalist forms of spatial organization, as well as building and urban forms were introduced. Where substantial indigenous urban systems existed as, for example, in North or East Africa, or South-East Asia, new urban hierarchies were introduced (Abu-Lughod, 1976; 1978; 1980; Soja and Weaver, 1976; Saueressig-Schreuder, 1986), frequently leading to megacephalous primate city growth, often in port cities.

Colonialism likewise affected urbanization and urban forms in the metropolitan country itself. Most significantly, each European empire had its own urban hierarchy, the imperial capital of which was both fed from the extraction of colonial surplus (cultural, as well as economic accumulation) and also bloated by the provision of financial, political, and administrative services for its colonies, and occupying the predominant place in the overall imperial urban hierarchy. As indicated in the companion volume (*Urbanism, Colonialism, and the World-Economy*), London in 1931 with a population of eight million, was almost six times larger than the next largest city in the Empire, Calcutta, an extreme example of metropolitan dominance (Christopher, 1988). The sizes of other metropolitan cities in the imperial urban hierarchy were, as discussed in more detail below on p. 36, equally bloated — Manchester, Liverpool, Glasgow.

In the process of incorporation, of different cultures into one world-economy, certain languages (and cultures) had a privileged status, particularly English, and also Spanish and French. In those areas influenced by 'Western' expansion or ideas, though not necessarily subject to formal colonialism, other languages and cultural influences were also important, particularly Arabic, Chinese, and Russian. However, in the nineteenth and especially the twentieth centuries, principally through colonialism and American economic and political hegemony, English as a world language and carrier of scientific and cultural practices has become a significant factor in creating the contemporary world system (McCrum, Cran, and MacNeil, 1986).

At each link in this urban chain, there was an ongoing process of economic and political, as well as social, cognitive, cultural, and environmental transformation, conceptualized during different periods by a variety of terms: during the period of colonization, 'Europeanization', 'Westernization' and 'cultural change'; from the mid-twentieth century, 'modernization', 'development', and later, 'bourgeoisification' and 'globalization'. All these terms, however, find spatial expression in the processes of urbanization and urban development and are often (literally) concretized in the built environment. (A proper understanding of the emergence and meaning of such terms awaits systematic investigation - see, for example, Wolff-Phillips, 1987, on the origins of the term 'Third World'.)

WORLD-CITY HIERARCHY

The question then arises as to which of these colonial towns and cities, including the metropolitan centres at the core, become incorporated into the world-city hierarchy, at what date and particularly, what criteria favour one place rather than another.

Whilst population size is not necessarily an indication of economic and political importance, Chase-Dunn (1985: 279) none the less provides an interesting insight in his table showing the world city-size hierarchy in the capitalist world-economy for the ten largest cities between 1500 and 1975. As might be expected for much of this period, these represent the economic and political strength of their respective states in the contemporary world-economy.

Between 1500 and 1700, the ten largest cities are all in Europe. In the mid-sixteenth century, they reflect the importance of the northern European-, Mediterranean- and emerging Iberian-based world-economy (Paris, Naples, Venice, Lyon, Granada, Seville, Milan, Lisbon, London, and Antwerp). By 1700, the ten largest cities were still in Europe and were principally the capitals of nation-states or early colonial empires (London, Paris, Lisbon, Amsterdam, Rome, and Madrid) as well as the Italian centres of the earlier Mediterranean era (Naples, Venice, Milan, and Palermo).

Between 1750 and 1850, regions drawn into the capitalist world-economy through major urban centres include Russia (Moscow), Turkey (Constantinople), Egypt (Cairo), India (Patna, Bombay), and China (Peking, Canton, Hangchow, Yedo, and Soochow).

However, the strength of present world cities of core countries, as measured by their place in the city-size hierarchy, was already evident by 1875, and clearly established by 1900 when the ten largest cities in the world were, in order of magnitude: London, New York, Paris, Berlin, Chicago, Philadelphia, Tokyo, Vienna, St Petersburg, and Manchester. Here, however, Britain's imperial role becomes clearly evident. The extent of the specialized industrial–commercial role of Britain in the world-economy, and especially as head of its imperial urban hierarchy, is measured not only by the size of London, but by the fact that *five* of the world's nineteen largest cities in 1900 were in the United Kingdom, each of them connected, not simply to the 'world-economy' in general, but specifically to a colonial mode of production — London, Manchester, Birmingham, Glasgow, and Liverpool (Chase-Dunn, 1985: 276). The size of Manchester, for example, must be seen in relation to the structural transformation of the Egyptian economy as it was integrated into the world capitalist system producing cotton as a consumption commodity, with the output of cotton increasing six times between 1860 and 1880. It must also take into account the increased foreign investment in Egypt and loans, 'often on highly unfavourable terms', with links to the London and Paris finance markets and the foreign banks in Egypt opened in 1880 (Chaichian, 1988: 28).

From 1920, Latin America (through Buenos Aires) moves into the league of continents with the world's ten largest cities and in 1940, China (Shanghai) and India (Calcutta) reappear. Between 1970 and 1975, the further urban incorporation of Latin America into the capitalist world-economy is manifest in the entry of Mexico City and São Paulo and also of Japan, through Osaka. In the mid-1980s, depending on definition, the world's ten largest cities in order of magnitude were Tokyo, New York, Mexico City, Osaka, São Paulo, Seoul, London, Calcutta, Buenos Aires, and Los Angeles (Feagin and Smith, 1987: 8). (It is worth mentioning that while Friedmann (1986) includes all of these, except for Osaka and Calcutta, in his world-city hierarchy, Feagin and Smith (1987) point out that São Paulo and Calcutta, both with over 11 million population, had no multinational headquarters in 1984 and Mexico City and Buenos Aires only had one each. All ten cities had, in Feagin and Smith's definition, populations over ten million.)

However, it is clear that the mere fact of size alone is insufficient to give world-city status: a variety of other factors are much more important, not least the strength of the economy to which the city belongs, its location in relation to zones of growth or stagnation in the international economy, its attraction as a potential basing point for international capital (for banks, multinational headquarters, and producer services), its political stability, and especially its historic and cultural connections to other world cities both in the semi-periphery and the core. For these reasons, Sydney, Singapore, and Hong Kong rank as world cities, whereas, with the exception of Johannesburg (whose future, with sanctions bringing economic and political instability, is uncertain) no African city has, by these criteria, been classed as a world city (see Simon, 1989).

In the following section we examine the role of colonialism in establishing the structures for the subsequent transformation of a colonial city into a world city.

WORLD CITIES, COLONIAL CITIES: COMPARISONS

Making comparisons between two ideal-type categories of the historical colonial city at the periphery, and the contemporary world city at the core of the capitalist world-economy, is both a theoretical as well as practical exercise. It also poses analytical problems: it could, for example, be argued that it would be more productive to compare today's world cities with today's ex-colonial cities, as implied by Armstrong and McGee (1985, Chapter 3). Yet, such a diachronic comparison over time makes sense when it is recognized that colonialism was one stage in the historical development of global capitalism, a process that expanded geographically from the capitalist core of Europe to other continents (King, 1989b).

In this context, in terms of the incorporation, through colonialism, of pre-capitalist, pre-industrial, and non-European societies into the world-economy, colonial cities can be viewed as *forerunners* of what the contemporary capitalist world city would eventually become. For it was in the colonial and paracolonial societies of especially Asia, Africa, and Latin America where the representatives and institutions of industrial capitalism first confronted those of ethnically, racially, and culturally different pre-industrial and pre-capitalist societies at any significant scale. This historically significant phenomenon, though charted in the scholarly literature in the last three decades has, as indicated in the earlier discussion of colonial cities (King, 1989b), been previously inadequately conceptualized and misunderstood, described as a problem of 'Westernization' or 'Europeanization', rather than the expansion of the capitalist world-economy. The institutions, as well as physical–spatial forms introduced in the colonial city, were not only 'European' but the 'norms and forms' of the capitalist industrial city.

From the nineteenth (and earlier) to the mid-twentieth century, the political and economic strength of core imperial powers kept this old international division of labour in place, and in the process, kept colonial peoples and institutions distanced from the core. Likewise, they safeguarded their own colonies as regions for capital investment from their own multinational companies.

The second half of the twentieth century, however, has seen the demise of colonial regimes. In three decades of post-imperialism following the war, over seventy once-colonial nations gained independence. Aided by the unfettered operations of the world market economy, the rise of global corporations, and revolutionary developments in transport technology, international labour migration has grown apace (Thrift, 1986b; Cohen, 1987). In this new situation, the site for the interaction, and confrontation, of representatives of underindustrialized and economically poor societies, and racially, ethnically, and culturally different representatives of advanced industrialized, and rich capitalist societies, once confined to the colonial periphery, has moved to world cities at the European and North American core.

In the ex-colonial societies, different political ideologies have produced different urban outcomes as Abu-Lughod (1984) shows in her discussion of the neocolonial urban policies of Tunisia and Morocco and the state socialist one of Algeria. Likewise, socialist regimes have been established in Vietnam (previously French Indo-China), Zimbabwe (Rhodesia), Mozambique, Cuba, and Nicaragua (Forbes and Thrift, 1987). Yet in many ex-colonial cities (including Singapore, Jakarta, Kuala Lumpur, Manila, as well as colonial Hong Kong (Drakakis-Smith, 1987)) there is increasing convergence between their characteristics and the forms and attributes of world cities at the core despite the considerable historic and cultural differences (Armstrong and McGee, 1985). Hence, while recognizing the variations in their histories, to make comparisons between the characteristics of historic colonial and world cities is both analytically legitimate and useful, particularly in providing criteria for evaluating the possible future development of other post-colonial cities, not least in Africa (Simon, 1989).

In the following section, therefore, which draws on the material in the chapter on colonial cities in King (1989b) (see Appendix 1), the objective is primarily speculative: to provide a set of ideas that might be tested by more detailed research.

Comparisons

1. The most important characteristic of both types of city is their external orientation: they are products more of external rather than internal forces, their fortunes dependent on decisions made out- side the state. This poses an inherent conflict of interest between the interests of external forces (whether the colonial power or transnational capital) and the national or local society. In both cases, considerable sectors of the city are externally or internationally rather than internally oriented.

2. As a result of this external orientation, the city becomes a major vehicle for political, cultural, and ideological transmission, an instrument for effecting economic and cultural change.

3. Both types of city are major communications and transport nodes.

4. Both colonial city and world city are major centres of political, economic, administrative, and managerial control, the former in relation to the colonial society, the latter in relation to the global economy.

5. Both types of city contain major financial institutions and are centres for directing international capital flows.

6. In terms of their economic and employment structure, there are broad structural similarities. Both types of city have a relatively large proportion of high-level administrative and producer-service class, that of the colonial city oriented towards political, administrative, and economic control, that of the world city to business, financial, and management control. In general, neither have heavy industrial employment: the world city at the core, because it sheds it, the colonial city, because it never developed it. Both have large middle-level and especially low-level service sectors.

7. In both the colonial and world city, a proportion of the elite derive their authority from external sources, in the colonial case, from the metropolitan power, in the case of the world city, from the corporate authority of transnational capital.

8. The link between internal and external forces (colonized and colonizer, national and international) gives rise to the development of a 'compradore' class, a set of cultural brokers, an international elite who hold privileged status in the society and make a significant impact on its social and political processes.

9. The colonial city as well as the world city are major destinations for national and international migrations of skilled, semi-skilled, and unskilled labour. These latter swell the ranks of the large tertiary sector in the informal economy, which is also characteristic of both types of city.

The population of colonial cities is formed, by definition, from international (i.e. colonizing) and indigenous inhabitants. In many cases, there is also what Horvath (1969) calls an 'intervening' or third-country group, for example, Indians in East and South Africa or the Chinese in Malaysia, such that the characteristic feature of many such cities is that of cultural and ethnic pluralism. Such pluralism, however, does not imply equality but rather unequal incorporation into the political economy and society.

With regard to core world cities, patterns of international migration are obviously not 'neutral' but take on the characteristics determined by various factors: their historical connections to previous colonies, employment potential, state control over migration, and cultural orientation; hence, the predominance of North Africans in Paris, Indonesians in Amsterdam, Goans in Lisbon, Latin Americans in

Madrid, Puerto Ricans and Philippinos (as well as other groups) in New York, and Indians, Pakistanis, Caribbeans, Australians, South, East, and West Africans, Canadians, and Americans in London. There are, of course, also groups from other countries of origin.

Given the growing importance of political struggles based on nationalist, ethnic, religious, or racial causes (for example, the support for Black South Africans among Afro-Caribbean populations, for ethnic or religious minorities in Sri Lanka or India, by co-religionist Sikhs in London or Toronto, or for nationalist causes, such as for the Irish in New York, or for the various religious and political movements in the Middle East), it is evident that the different national, religious, ethnic, racial, and cultural composition of world-city populations could have at least as much importance as their mere economic value as labour power in the world city. This is particularly so in determining its level of social and political stability in relation to threats posed by powerful actors on the world-economic and political scene (in this context, the presence or absence of particular groups of Sikhs, Iranians, Nigerians, or Americans in London have had important economic and political consequences in recent years).

In brief, political and social threats to stability from conflicts amongst racial, ethnic, religious, or national minorities are equally important as economic threats. In this way, Beirut has lost its potential world-city status since the early 1970s (Abu-Lughod, 1984; King, 1988). As well as political regimes, the population composition of any city, resulting from past migrations and the potential for conflict that they bring, will influence the likelihood of any city in the periphery or semi-periphery to take on world-city status, whether in terms of attracting investment, employment or as a host for global events (cf. the recent histories of Colombo, Kuala Lumpur, Seoul, and Johannesburg).

10. In both types of city, to varying degrees, occupational stratification is by ethnic and racial group. Regarding this social, ethnic, and racial composition, non-indigenous or non-national residents inhabit a disproportionately large number of the dominant positions in the social structure (the colonial elite or representatives of transnational capital) and a much larger proportion of subordinate positions, either as low-income country migrants in the world city or 'third-country' migrants in the colonial city. The middle ranks of society are largely occupied by indigenous inhabitants.

41

11. Both types of city exhibit economic, occupational, social, and racial polarization. In the world city, this results from the operations of the market, the homogenization of economic space at a global level, with wages rising and falling in relation to labour supply, though as affected by institutionalized racism (Wallerstein, 1983: 78) and other factors. In the colonial city, this results from the confrontation between representatives of industrial-capitalist societies, as colonizers, and representatives of pre- or little industrialized societies and the institutionalized racism and segregation of colonial rule.

12. In both types there is a high degree of economic and social polarization between indigenous and exogenous inhabitants. In general terms, in colonial cities, exogenous groups are in superordinate positions and indigenes in subordinate ones; in the world city, this situation is reversed.

13. In both types of city, to varying degrees, economic, social, ethnic, and racial polarization result in spatial and residential polarization, as well as ghettoization.

Here, some comment is needed on the metropolitan and colonial cities in the pre-independence era. In the core metropolitan cities (London, Paris, and New York), class polarization was largely a by-product of market relations and resulted from economic and ethnic disadvantage and racial discrimination in varying forms in the democratic state. However, because of the position of core societies in the international division of labour and access to the global surplus, class and spatial polarization was (compared to pre-industrial experience of their own societies or equally, to the pre-industrial experience of the colonial societies), less extreme; in a number of cases, it was also alleviated by state intervention and the provision of welfare benefits. Class polarization, in brief, only had a national dimension.

In colonial cities, the situation was quite different. First, class and racial polarization, spatially expressed, was already at a global or international scale as a result of the place of the colonized society in the international division of labour. Because of the appropriation and redistribution of the surplus in colonial societies, colonial (i.e. European) populations lived at, or usually higher than, the standard of living in the metropolitan society; indigenous populations lived at, or below the standard of living of the colonial peasant society. Moreover, state intervention to provide welfare benefits (housing, health, and education) to the indigenous population was, in general, either very limited or non-existent.

In short, social and spatial polarization was not a product or accident of market forces but, in most cases, a result of colonial policy, which also ensured strict segregation between colonial and colonized populations (the circumstances varied according to both the particular colony and period under discussion).

Hence, what Friedmann calls 'the major contradictions of industrial capitalism, among them, spatial and class polarization' (Friedmann, 1986: 76) expressed in extreme, and also racial form, were evident in colonial cities long before they became evident on a large scale in core world cities, although observers in such world cities may have taken little notice of these phenomena (colonial subjects were in core cities before independence though their numbers were relatively few).

The difference is that, in the last two or three decades, developments have occurred to reproduce, in some of today's core world cities (such as New York or Los Angeles) the degree of racial, class, and spatial polarization previously characteristic of peripheral and semi-peripheral colonial cities.

Racial and class polarization on a global scale, mediated to varying degrees by state policies (whether in regard to immigration control, welfare provision or planning strategies), once confined to the historic colonial city, is now produced in core world cities. Whilst some observers (e.g. Ross and Trachte, 1983; Mitter, 1986a) see comparable standards of deprivation between parts of core world cities and urban conditions in the semi-periphery, showing, for example, similar mortality and morbidity statistics in immigrant ghettoes in New York and cities in Latin America, it is clear that, at present, conditions in core cities are less extreme than those in the periphery. However, such a generalization would need testing by reference to individual cases (see, for example, Soja *et al.*, 1983; Armstrong and McGee, 1985). At the top end of the social hierarchy, however, there is plentiful evidence that life styles become increasingly comparable. Should market-force ideologies prevail on a global scale, and with little intervention by individual states, the disparity between degrees of polarization in cities within the capitalist world economy could, theoretically, even out. The main controlling factors will be the fortunes of the individual state in the larger world-economy and the policies that it adopts.

14. Both types of city provide the site for the confrontation and encounter between representatives and institutions of different nations, ethnicities, races, religions, and cultures. Both types of city provide opportunities, though in markedly different economic and political contexts, for the redefinition, transformation, and reconstitution of social, political, cultural, ethnic, and biological categories.

15. Both world cities and colonial cities are sites for the concentration and accumulation of international capital.

This particular characteristic highlights different roles played by cities at various stages in the accumulation process. In recent years particularly, major world cities (Los Angeles, New York, London, Sydney, Singapore, and Hong Kong) have become major centres for investment in commercial and domestic property (Soja *et al.*, 1983; Thrift, 1986a). At an earlier historical stage, in the periphery and semi-periphery, city- and settlement-forming processes provided significant opportunities for the concentration of international capital, often in countries with few or no urban settlements, and involving the appropriation of land, its subdivision and the creation of a market.

This process is important in three respects. First, in commodifying land and the built environment generally, and creating a market where, in some cases, it had not existed before; second, in establishing cities as sites for the concentration of capital within the colonial system; and third, for providing, through this colonial system, the infrastructure for the present world-urban system now in operation.

Whilst some of these processes have been investigated (e.g. Hopkins, 1980; Abu-Lughod, 1980; Fetter, 1976; Simon, 1989; Ribeiro, 1989) there is need for much more systematic research.

16. Because both types of city fulfil powerful political, economic, and administrative functions, the extent and responsibility for which extends beyond the boundaries of the society where they exist, the powers of local government are likely to be subordinated to those of the national centre. In colonial cities, effective local government is suppressed; in world cities, it is likely to be subordinated. Both for these reasons, as well as the potential for political and social conflict, institutions of social control as well as expenditure on policing are proportionately greater than in other urban centres.

17. In both types of city, state spending favours external interests. In the colonial city this is at the expense of the local population.

In the colonial city, state and municipal spending favours the colonial elite; in the world city, state spending in the public domain (through tax concessions, enterprise, or free-trade zones) increasingly favours transnational capital.

18. Both types of city are ideological centres, one in relation to the colonial state, the other, in relation to the state and wider region.
19. Both world cities and colonial cities generate social costs at rates that tend to exceed the fiscal capacity of the state.

In the case of both types of city, supranational functions are, to different degrees, subsidized by national resources. Where, in the world city, they are financed by international capital, they benefit the international elite. However, the rapid influx of poor national and international workers into world cities, according to Friedmann (1986), generates massive needs for social reproduction, particularly housing, education, health, transportation, and welfare. These needs are increasingly arrayed against other needs arising from transnational capital for economic infrastructure and from dominant elites for their own social reproduction. In this competitive struggle, the poor and especially the immigrant populations lose out.

This characteristic has much in common with the colonial, and especially post-colonial city. Originally imposed from 'outside', it was sustained by surplus extracted from the colonial economy. With the demise of the colonial power, it lacked the economic base to meet the social costs it generated. In comparison with urban centres in industrial societies, the ex-colonial city was said (in the early 1970s) to have 'problems of housing, a shortage of economic resources, underdeveloped communication systems and a lack of the institutional infrastructure to deal with social, administrative and political needs' (King, 1976: 23). It is a reflection on the homogenization of global economic space in the last fifteen years that these comments are increasingly relevant to the world city in the core.

Whilst these comparisons have emphasized the similarities, the differences are of equal importance. Of these, the most significant factor is the most obvious one: the lack of economic and political independence that defines the dependent nature of the colonial state. It is the degree to which states can and do exercise that independence that post-colonial cities will differ — both from world cities at the core as well as from each other.

CONCLUSION

What these comparisons suggest is that structurally, colonialism — as a form of political economy or, in Abu-Lughod's (1984) terms, a historical 'mode of production', sets up much of the infrastructure, whether this is economic, political, social, cultural, as well as physical and spatial, for the later world city: it is a process of 'softening up', a 'modernization' for transformation to world-city status. This is true for the capital in the colonial society, the major, colonial port city or the metropolitan capital of the colonial empire. Colonialism establishes the institutions, the banks, corporations, governments, commercial forms, hotels, as well as the educated elites, consumption habits, retailing behaviour, technology, attitudes, and aspirations on which global capitalism, and subsequently, the world city builds. It is this that has turned Asian, Latin American, and, to a lesser extent, African cities into real or potential instruments for world-market expansion; yet critically, of course, only in those states that have remained committed to the world-market system.

In the fifteen years that have elapsed between the setting out of a tentative theory of colonial urban development (King, 1972, revised 1976), many of the distinctions between present-day world cities in the core and ex-colonial cities on the periphery or semi-periphery have disappeared. The removal of particular colonialisms from the 1960s (French, British, and Portuguese) and the general expansion of the world market, though characterized by grossly uneven development, has none the less evened out some of the differences by two processes: the peripheralization of the core and the corealization of the periphery. It is appropriate, therefore, to conclude this section with some comments on 'the colonial city transformed'. Whilst these are brief and essentially impressionistic, they provide some directions for further research.

The colonial city transformed

In the late 1960s, twenty years after Indian Independence in 1947, the Indian capital of New Delhi still had many of the characteristics of the archetypical colonial capital city. Despite massive urbanization encircling and enlarging the Old Delhi–New Delhi area, the colonial capital built between 1911 and 1940 was still very much in place: the huge Government House (now Rashtrapati Bhavan) looking down the vast expanse of the Central Vista (now Rajpath), the two large Secretariat buildings, the grand tree-lined avenues dividing up the hexagonally arranged areas of residential space in which hundreds of individual official bungalows, each located in generous three-, four-, or five-acre compounds, expressed in size, location, and distance from Government House the status of their original occupants. At the northern edge of the city, acting as a pivot between Old and New Delhi and articulating the economic and social relations between them, was the spacious, two-storey retailing centre of Connaught Circus (King, 1976, Chapters 1, 10).

In terms of our earlier explanations, this was a built environment produced by a colonial mode of production, one that had its counterparts in other colonial territories such as French North Africa or British South-East Asia. Yet to understand its distinctive forms we need to recognize that the colonial inputs were culturally British just as the colonized society was culturally Indian, though each of these broad categories covering a wide range of social, religious, and regional subcultures and variations. And having recognized these structural levels of explanation, we can see it as the product of particular agents representative of British imperial power: the architect-planners Lutyens and Baker as well as their British and Indian assistants.

In 1967, however, India was a democratic republic with a mode of production based on the public ownership of large sectors of the economy though with a growing private sector. Technically, it was a 'mixed economy'. In this context, the capital of Delhi largely continued its colonial or imperial function as a centre of government and administrative control. There was relatively little industry but extensive public-service employment in government, education, health, scientific and medical research, and cultural institutions and this was reproduced in the buildings and residential environment. A large part of the land was government-owned and controlled.

Hence, much of the colonial-built environment remained intact: the huge bungalows continued to be occupied by government ministers, senior administrators, and the high-ranking military, judicial, commercial, or foreign diplomatic elite.

Twenty years later (1987), the centre has been transformed, taking on many of the attributes of the modern capitalist city. Whilst the

47

mode of production is still that of a mixed economy with democratic control, under the more market-oriented policies of the present government, the balance of that mix has changed: more space has been given to the private sector and new external forces have impinged on both the economic and physical territory of the state. American, Japanese, West German, and British multinationals all have operations in India. Delhi is increasingly a regional centre, though subordinate to Bombay, in the capitalist world-economy, and this is expressed in the built environment.

The old colonial retailing centre, the two-storey collonnaded circle of Connaught Circus, is now ringed by massive high-rise office blocks, a phenomenon that began at the end of the 1960s. Whilst banking contributes a significant number of tenants, the growth of branches of foreign banks in Delhi (as opposed to Bombay) has been relatively modest. Where there were four in 1970, there are ten in 1987 though it is some of the major ones (the Bank of America, the Banque Nationale de Paris, and Citibank) that are the principal tenants of the high-rise office towers, one of the highest characteristically named 'The New Bank Block'.

The new Stock Exchange, rising to some thirty storeys was opened in 1985 in the main financial centre in Bombay, which contained nineteen of the twenty-one foreign banks' regional offices. It was designed by a consulting engineer, graduate of the University of Bombay and Associate Member of the Institute of Structural Engineers (London). (See *Architecture and Design. A Journal for the Indian Architect* (1985): 1, 6; see also Appendix 2 (p. 159) on foreign banks in India.)

One of the largest blocks in Delhi houses the State Bank of India. In the early 1980s, this had forty-four overseas offices in twenty-eight countries and in 1983, opened new offices in Sri Lanka, Toronto, and Los Angeles, as well as others in the Gulf, mainly to finance joint ventures abroad. In 1981–2, advances of the bank rose by 65 per cent, mainly accounted for by the Bahrain, London, Singapore, and Tokyo offices. In the United States, the Bank's New York branch was financing hotel development in Daytona Beach, Florida, whilst the Los Angeles branch had invested Indian funds in the prestigious Crocker Tower in Los Angeles. These, of course, were only part of a policy that included extensive loans to industrial and craft projects 'back home'. Yet highly competitive banking (including breaking conventional quarter-per-cent barriers and giving three-sixteenths of a per cent over prevailing rates) was increasing the Bank's business in South-East Asia (*Annual Report*, State Bank of India, 1982; 1983).

Other international companies have tapped India's underutilized and comparatively low-paid supplies of educated manpower. Both in Bombay and Delhi, IBM have set up offices to process software brought in from abroad (Noyelle, 1986).

Redevelopment in Delhi has included huge luxury hotels and super-market developments. Whilst the old 'Empire Stores' remains, versions of McDonalds have also made an appearance. Under the 'Emergency' regime of Mrs Gandhi, the Government adopted wide powers to prepare for Delhi's hosting of the Asian Games. In addition to a large new stadium, huge flyovers were built and a vast road-widening and motorway system was undertaken, linking Palam airport to the centre, in many places with six-lane highways. For considerable parts of the day, these are mainly used by motor scooters. The impact on foreign visitors (whose numbers have increased fivefold from the 200,000 in 1970) is initially impressive, not least because the speed at which taxis now travel helps to conceal the *jhuggies* of poverty-stricken migrants knitted into the fabric of the city.

Only from the 1980s has the colonial residential centre become seriously affected, occupied as this was (and still is) by ministers, senior officials, and business elites. Whilst some of these low-density bungalow areas (now becoming part of the central business district) are being redeveloped, such was the degree to which values are shared between the ex-metropolitan (British) and ex-colonial (Indian) elites, that in 1974 a large section of the colonial residential area was designated (again, under legislative and institutional arrangements imported from outside) as a 'conservation area' under the authority of the Delhi Urban Arts Commission.

In other words, senior decision-makers in Delhi (including urban planners) made a reading of Delhi's physical and spatial environment that privileged the role of colonial architect-planners (Irving, 1983) over other readings that emphasized the inequalities of colonial urban development (King, 1976).

More recently, however, the debate has become more complex. According to a report in the *Indian Express* (21 January, 1987), 'the sprawling government bungalows in the capital's VIP zone are likely to give way to smaller, more compact modern houses and flats in the near future'. The Ministry of Urban Development plans to redesign the layout of 2,000 acres south of Rajpath (including Zones D 11 and D 12). According to figures from the last census, the population density of this bungalow area was still only 20 to 25 persons per acre which, under the Draft Master Plan for the year 2000, would be increased to 81 p.p.a. and would thus provide much needed housing for government employees. 'The bungalow concept would have to go' as plans are developed to 'carve up the enormous colonial residences set in 3 or 4 acres' and replace them with half-acre plots and duplex houses 'for ministers and other senior leaders taking into account their need for space for morning durbars and receptions'. Local Indian capital also presses for (and obtains) redevelopment. Simultaneously, ex-metropolitan interests in Britain support Indian pressure groups to conserve what they refer to as 'Lutyen's Delhi' in the (stated) interests of tourism and 'cultural heritage' and the (unstated) interests of cultural imperialism (Cruickshank, 1987).

In the rapidly growing suburbs of New Delhi, however (a city nearing six million compared to the 3.5 million in 1970), convergence between the one-time colonial and one-time metropolitan cities, between 'Third-World' and 'First-World' environments, now takes place. Whatever the fate of the growing numbers of the poor, elite life styles become increasingly similar (cf. Armstrong and McGee, 1985: 48–9).

Eleven minutes' drive from Delhi's international airport, Palam Vihar is developing, according to its promoters, as 'the elite neighbourhood of Delhi' and 'one of the most prestigious localities'. Designed for 50,000 residents it has 'every conceivable modern facility like schools, clubs, and an exclusive shopping centre'. Here, are offered 'a remarkable range of houses' including 'the Continental Villa', 'the French Villa', and various styles of 'cottages'. The Continental Villa has 'a European touch of class to it, including a lovely landscaped garden'; the French Villa, 'sloping red roofs, beautiful bay windows, quaint balconies' and 'a French flavour in every detail'.

What were once thought of as deep-seated cultural taboos (such as bathing in one's own dirty bathwater in Western-style tub baths or the use of Western-style closets (King, 1984a: 62)) are being abandoned. Three-piece bathroom suites are now manufactured not only with tub baths and upright (rather than 'Indian style') closets but also marketed (by Hindustan Sanitaryware and others) in a range of thirty different colours to provide 'a touch of Italian romance in your bathroom'. These bathrooms, kitchens, and drawing rooms are indistinguishable from others in the market economies of the Western world (*Architecture and Design*, 1, 6, 1985). Cleanliness and hygiene, once thought of in India in religious and ritual terms (Carstairs, 1968; Douglas, 1978; Kira, 1976) are commodified just like anything else.

The archetypical colonial city is being gradually transformed into the archetypical city of capital.

World cities of the future?

The basic assumption behind the world-city hierarchy is that global control centres are located in the core countries of the capitalist world-economy (Cohen, 1981; Feagin and Smith, 1987). Only the zonal or regional centres (in Friedmann's (1986) terms, the third- and fourth-order cities) are in countries in the semi-periphery (the middle-income countries according to World Bank criteria).

Which cities outside the core are likely to strengthen their status as 'basing points' for capital in the future, say, the year 2000, alongside the other 'important centres of capital accumulation' (Friedmann, 1986: 72)? And more importantly, what criteria can be used to address this question?

Given the historical expansion of global capitalism (and failing a world socialist revolution) the determining factor is, of course, the policies of individual states. Whilst changes may well occur in the industrial and post-industrial economies presently at the core, the future expansion of capitalist-based urban growth is clearly taking place in the industrializing and developing countries, both in the World Bank's 'middle-income' as well as, to a lesser degree, in the 'low-income' countries.

In this case, the criteria for future third- or fourth-order world cities might be listed as:

1. State policies oriented towards, or not opposed to market-oriented growth.
2. A minimum country population size of 15 million providing potential market growth.
3. The largest city in the country, with a minimum size of 1 million providing a potential market for consumer goods. (In certain cases, e.g. Kuala Lumpur, the size of the largest city may fall below 1 million population.) In the case of India (population 765 million) and China (population 1,040 million) in 1987, more than one world city is likely to develop.
4. A continuous average growth rate, 1965–85, in the region of 2 per cent.

If recent liberalization policies in China can be taken note of (see Western, 1985) these criteria would give the list in Table 3.1. This relates only to the World Bank's 'middle-income' and 'low-income' countries and does not include 'industrial market economies' in Europe.

It is worth noting that, with the exception of Istanbul and Tehran, all of these cities, currently national or regional centres, were either strongly influenced by Western colonialism, were direct colonial products or, in the case of China, colonial 'free ports'. On the basis of the discussion in this chapter, this historical factor is likely to be a significant criterion in determining their future role in the world-economy.[1] And the data lend broad support to the 'Warren thesis': that imperialism is the pioneer of capitalism (Warren, 1980).

Table 3.1 World cities of the future? (population in millions)

Asia	China	Shanghai	11.8
		Tientsin	7.7
S.E. Asia	Indonesia	Jakarta	6.5
	Malaysia	Kuala Lumpur	0.9
S. Asia	India	Calcutta	9.1
		Bombay	8.2
		Delhi	5.7
	Pakistan	Karachi	5.1
	Sri Lanka	Colombo	1.4
Middle East	Turkey	Istanbul	4.0
	Iran**	Tehran	4.5
N. Africa	Egypt	Cairo	5.0
	Morocco	Casablanca	2.4
Sub-Saharan	Nigeria	Lagos	1.5
Africa	Kenya	Nairobi	1.0
S. America	Colombia	Bogota	4.0
	Peru*	Lima	4.6

Source: United Nations, 1986; World Bank, 1987.
Note: * below 2 per cent growth rate, 1965–85 ** Whilst Iran/Tehran meet the 2nd, 3rd, and 4th criteria, present circumstances in regard to the 1st mean that it must be treated as a special case.

Chapter Four

WORLD CITIES AND COLONIAL URBAN DEVELOPMENT

Hypotheses and theories

INTRODUCTION

In 'the world-city hypothesis', Friedmann's main purpose is 'to state the main theses that link urbanization processes to global economic forces'; the hypothesis is about 'the spatial organization of the new international division of labour' (Friedmann, 1986). Friedmann makes clear that the hypothesis, consisting of seven 'loosely joined statements', is neither a theory nor a universal generalization about cities but a starting point for political enquiry 'primarily intended as a framework for research' (1986: 69); likewise, earlier formulations about colonial urban development theory were at best tentative (King, 1976, Chapter 2). None the less, examining each of these pre-theoretical formulations together provides some heuristic insights for furthering Friedmann's main purpose.

Whatever its merits, however, Friedmann's (1986) hypothesis does not link urbanization processes *in general* to global economic forces but only those aspects of them that relate to world cities: nothing is said about other parts of national urban systems with which world cities are connected and which are equally subject to global economic forces (conceptually, this problem is more effectively addressed by Moulaert and Salinas, 1983; see also Alger, 1988); it addresses only one part (albeit an important one) of the spatial organization of the new international division of labour. Here, it is worth looking at colonial urban development (CUD) theory, which differs from the world-city hypothesis (WCH) in three ways. The first concerns the formulation of the problem, the second and third, the spatial scope of the analysis.

CUD theory examines the impact of colonialism, as a particular form of political economy (or mode of production) on forms of urbanization, urbanism, and urban development in three spheres: the colonized society, the metropolitan society, and other regions of the colonial system (King, 1976). The WCH, as indicated above, deals with the impact of the capitalist world-economy only on specific cities (though Friedmann (1986) defines these to incorporate the notion of a metropolitan region).

This highlights the second difference, which concerns the larger economic, political, and spatial unit in which each set of urban phenomena is embedded: the WCH relates to the capitalist world-economy, 'the global system of markets for capital, labour, and commodities', the theory of CUD to the more spatially restricted, colonial political economy defined according to particular historical, political, and geographical criteria (though it is important to recognize that colonial empires were subsystems of the world-economy).

What is common to both formulations, however, is that each posits a systemic connection between urbanization and global economic (and for CUD, social, political, and cultural) processes; both assume that the function of one part of the urban system is comprehensible only by reference to the rest of it.

Third, a major concern of the WCH is with the contradiction of interest posed by the global functions of world cities and the interests of local and national communities. Likewise, other work on the world-city phenomenon (Cohen, 1981; Noyelle and Stanback, 1984; Sassen-Koob, 1984; 1985; 1986; 1987; Smith and Feagin, 1987; Timberlake, 1985) generally focuses on the way national urban hierarchies are being restructured as a result of global forces and world-city formation. Less attention has been addressed to the new global hierarchy, not only of world cities but of other urban agglomerations, which are both products and instruments of the new international division of labour. The phenomenon is best expressed in a comment of Dieter Lapple (1982): 'São Paulo is Germany's biggest industrial city'; in the 1980s, we are looking at the economic, social, and urban-built environmental expression of an increasingly integrated (and disintegrated) global space economy — though that is not to undermine one of the main arguments of *Urbanism, Colonialism, and the World-Economy* (King, 1989b) namely, the extent to which previous environments were *equally* a product of an older international division of labour.

Here, it would again be useful to look at CUD theory. This, with revisions, provides a spatial framework for examining the impact of colonialism on urbanization and urban development in both the metropolitan and colonial society but also, by analogy, of the *contemporary* impact of global capitalism on core and peripheral countries. In doing this, it also provides insights into the historical continuities (and discontinuities) between the effects of colonial and subsequent world-economy processes.

WORLD-CITY FORMATION AND
COLONIAL URBAN DEVELOPMENT

We can start by reaffirming an earlier proposition: that in the words of Walton (1984: 78) particular 'modes of socioeconomic organization and political control' lead to particular forms of urbanization and urbanism and this also includes distinctive types of building, architectural, and urban form. Abu-Lughod (1984) also demonstrates the same principle, though using the briefer 'mode of production'. Using data from different countries in the Arab world, she shows how very different types of urbanization have resulted from different modes of production and political ideologies that she categorizes as 'neocolonial', 'state socialist', 'oil and sand', and 'charity cases'.

In the following discussion, attention is focused on two different though historically and structurally related systems of socioeconomic organization and political control or modes of production. The first is modern industrial colonialism of the nineteenth and first half of the twentieth century, an intrinsic part of the old international division of labour. The second is the contemporary capitalist world-economy of the late-twentieth century, which expresses itself as a new international division of labour. The illustrations are drawn from the experience of Britain, the Empire and the world-economy.

In each of these two cases, the mode of production makes a particular and distinctive impact at either 'end' of the spatial division of labour. Colonialism produces particular patterns of urbanization and urban development in the colonial periphery but likewise, it makes an impact on urbanization and urbanism at the metropolitan core. Moreover, as we have indicated earlier, as we are dealing with a *single* economic, social, and environmental system, these phenomena are not simply interrelated but are also complementary. Developments that happen at the core do not occur at the periphery and vice versa (i.e. The Colonial Office is in London and not in the colonial capital; industrialization is in metropolitan regional cities and not in colonial towns; the plantations are in the colonies and not in the metropolitan economy).

55

Obviously, colonialism was not the only factor influencing British urbanization and urban development in the old international division of labour: the degree and forms of these also resulted from national and international economic forces that, to varying extents, were independent of colonialism. We can, for example, ask the counter-factual question: how would Britain's economy — and urban system — have developed without the empire and colonies?

Yet the assumption behind the methodological framework proposed here is that the privileged position derived from political control over markets, sources of raw material, and labour inherent in Britain's imperial status was a major factor influencing urban and regional development in the UK. It has already been stated that in the 1930s, the colonies and dominions took over two-thirds in value of goods exported, whilst something under half, in value, of goods imported came from there. These proportions also changed over time. More important was the fact that they affected some regions, towns, built environments, populations, and urban social structures much more than others. How else explain the bizarre phenomenon of the Lancashire 'cotton towns' (sic)?

Table 4.1 A conceptual framework to examine the influence on urbanization and urban development of colonialism and global capitalism

Influence on urbanization and urban development of			
Colonialism *(19th to mid-20th centuries)* on		*Global capitalism* *(mid-20th century —)* on	
1	2	3	4
Colonial society	Metropolitan society	Ex-colonial society	Ex-metropolitan society

The first case, the impact of colonialism on urbanization and urban development in the colonial society was examined in *Colonial Urban Development* (King, 1976), one aim of which was to suggest a methodology for understanding the issues discussed. In that account, a spatial framework was suggested with which to examine the impact of colonialism at a number of different levels i.e. global, international, national, urban, sector, and unit (see Table 4.2). At each of these different levels, attention may be focused on economic, social, and cultural aspects of the problem investigated. In the original study, the emphasis was on social and cultural aspects and the way in which these were reproduced in physical and spatial form.

This framework can also be used to examine the influence of colonialism on urbanization and urban development in the metropolitan society (only briefly outlined in the original account) as well as global capitalism in both places. However, as no study has, as yet, been made of the way in which recent world-economic forces have affected Indian urbanization and urban development (the case study originally examined — though see Chapter 3, pages 46–51), column 3 in Table 4.1 cannot be completed.

The following account focuses only on the metropolitan society, and uses a selection of examples to illustrate how the impact on urbanization of these two historical modes of production might be examined.

The framework, therefore (see Table 4.3), is primarily a *method* with which to examine the links between 'urbanization and global economic forces'. The various levels do not provide an *explanation* but rather, act as a heuristic device to unpack the other elements in the larger space economy, whether these are in the national or international sphere. How valuable the framework is will depend on the extent to which it helps to provide explanation. As the detailed description of the different levels is given in *Colonial Urban Development* (King, 1976), only brief reference is made to them here. The examples also will focus more on built-environment phenomena though the framework can also be used to draw attention to economic, social, cultural, or other aspects.

The global level

As originally formulated (King, 1976: 26–8), this level of analysis concerns cognitive phenomena and, in the metropolitan society, covers the processes by which knowledge and assumptions about urban systems in the colonial and ex-colonial societies are derived by members of the metropolitan. (In the colonial society, it concerns the transfer of urban-planning theories from the metropolitan core to the colonial periphery, i.e. theories derived from one mode of production and transplanted to the territory of another.)

Under contemporary global capitalism, this level covers, in the academic and professional sphere, knowledge and theory about, for example, world urban systems, world cities, globalization, global restructuring, i.e. knowledge that may influence decisions about urban development and planning. In the popular sphere it refers to knowledge

Table 4.2 A conceptual framework for the study of colonial urbanization (colonial society: India)

Scale of geographic unit	Dimension	Scale of social unit	Type of phenomena considered	Illustration
1. Global	Macro	Intercultural	Cognitive	Knowledge of urban systems, planning theory
2. International or 'imperial'	Sub-macro	Intersocietal	'Centre-periphery', 'cultural pluralism'	'Third country urbanization and urban development' (in other societies of the colonial empire)
3. National	Major	Societal	Economic, organizational, spatial, cultural	Seaports, district towns, colonial capital, 'cantonments', 'hill stations', considered as a *system*
4. Urban	Intermediate	Urban	Social, spatial, cultural, economic, technological	'Colonial city' structure: 'native city'/'cantonment'/ 'civil station'
5. Sector	Minor	Community	Social, spatial, cultural, economic	'Colonial urban settlement' structure: 'indigenous city' structure: 'cantonment structure'
6. Unit	Micro	Institutional (domestic or public)	Built form (architectural), social, cognitive, cultural, economic	Residential units: 'bungalow-compound', courtyard-house, church, mosque, temple, barracks

Source: King, 1976: 27.

Table 4.3 A conceptual framework to examine the influence on urbanization and urban development of (i) colonialism and (ii) contemporary global capitalism (metropolitan society: UK)

Scale of geographic unit	Colonialism (old international division of labour) (19th to mid-20th centuries)	Global capitalism (new international division of labour) from mid-20th century onwards
1. Global	Communication and cognitive system operating at level of ideas	Technologically advanced communication system/cognitive system, etc. (also 'global trading', etc.)
2. Imperial/ international	Metropolitan capital considered in role of 'command centre' of imperial and colonial system (London as 'imperial city')	Transformation of imperial city into world city, with specific functions in capitalist world-economy
3. National	National structure of urbanization and urban development as influenced by colonial mode of production (e.g. regions/localities largely dependent on colonial connection, especially ports, manufacturing areas, residential locations)	National structure of urbanization and urban development (urban and regional system) as transformed by contemporary global capitalism and role of state in new international division of labour
4. Urban	Economic, social, and spatial structure of cities as influenced by colonial mode of production. Includes both national capital (London) and regional cities	Urban economic, social, and spatial structure as transformed by contemporary global capitalism and by new roles in capitalist world-economy
5. Sector	Particular sections of city determined by or dependent on colonial connections, whether economic, social, residential, etc.	Sectors of city (whether capital/world city or regional city) as influenced by contemporary global capitalism
6. Unit	Specific institutions and built forms/types dependent on colonial connection	Specific institutions and built forms/types as influenced and transformed by contemporary global capitalism and new roles in capitalist world-economy

Source: King, 1976: 27

of Third- or First-World cities used by the inhabitants of each to make comparisons between them. This level is also appropriate for discussing global electronic communications, investment flows, etc.

The international level

Imperialism and colonialism[1] are manifest in innumerable forms in the metropolitan capital. The imperial and colonial control functions (to use Friedmann's (1986) phrase) are not only 'directly reflected in the structure and dynamics of their production sectors and employment' but also in the city's huge size, institutions, economic, social, and cultural composition as well as in its built environment.

Whilst this is discussed at greater length in Chapter 5 on London, mention may be made here of those particular attributes of the metropolitan capital resulting directly from its role of control, i.e. institutions of government (the Colonial Office, India Office, and the Foreign and (especially) Commonwealth Office); financial and commercial sectors that benefit disproportionately from the international investment and related functions of the colonial connection (banks, insurance, shipping offices, docks, and many related industries (West India Docks, East India Docks, etc.)); commercial and residential property developed from colonially derived investments and surpluses; retailing establishments dealing in colonial goods or catering for colonial populations (Whiteley's, Army and Navy Stores, etc.); shipping, travel and despatch agencies; a variety of educational, scientific, and cultural institutions associated with the appropriation, development, and reconstitution of colonial cultures and societies (e.g. Imperial College of Science and Technology, the School of Oriental and African Studies, the London School of Hygiene and Tropical Medicine, the Tropical Products Institute, India Office Library, the Royal Colonial Institute, the Royal Asiatic Society, various missionary and religious headquarters, particular museums and research institutions, 'dominion agencies' for Australia, Canada, New Zealand, the Crown Colonies, the South African Colonies, the Malay Straits, etc., for controlling the flow of international labour migration and settlement (as well as offices of the Canadian Pacific Railway).

Such institutions (and economic, social, and occupational structures that relate to them) become especially evident when the city is compared either to imperial capitals with similar functions (e.g. Paris and Amsterdam), to colonial capitals initially without such institutions

(Delhi and Algiers), to other metropolitan cities with mainly commercial or industrial functions (e.g. Manchester and Birmingham) or to other non-colonial national capitals (e.g. Oslo and Athens).

In sum, everything concerning the size, population, institutions, and requirements of an imperial capital are inflated far in excess of what the metropolitan society on its own, either requires or, without the tribute from the empire, can economically support. The success of economic colonization inflates in particular the monetary economy of the metropolitan capital, and hence, its banking function, far beyond the requirements of the local population.

Whilst such institutions have their historic origins in colonialism, they are frequently the antecedents of contemporary internationally directed and 'development'-oriented activities. They are a prerequisite for the development of the institutional and professional infrastructure of the world city and are critical for establishing its ideological and cultural role.

This level of analysis, then, deals with the integrated ensemble of functions, activities, and institutions determined by the role of the city in the prevailing mode of production. Studies of particular sectors or institutions in the city are referred to on pp. 64–6.

In examining the impact of global capitalism on urbanization and urban development, this is obviously the level of world-city analysis. This focuses both on the production sectors and employment as well as their functional institutions that, while present in core world cities (Hall, 1984) are not located elsewhere in the cities of the periphery or semi-periphery (Cohen, 1981): the multinational headquarters. offices, international institutions, international banks, commodity and stock exchanges, producer-service concentrations, cultural- and media-production centres, as well as the ancillary services dependent on them such as hotels, restaurants, the police and security services, entertainment, etc.

The national (and regional) levels

This level focuses on the national urban system, the characteristics of urbanization and urban development as influenced by a colonial mode of production in the nineteenth and early-twentieth centuries and global capitalism in the late-twentieth century.

With regard to colonialism, the original formulation (King, 1976: 29) referred to particular urban areas, such as industrial regions and ports (e.g. Glasgow and Liverpool), primarily dependent on the import and manufacture of colonially derived raw materials; land and property development resulting from the investment of resources derived from colonial trading; and in terms of residential use, selected towns and semi-rural areas (e.g. resorts and spas and the 'Home Counties') preferred as residential locations by returned colonial ex-patriates, or benefitting by their patronage (i.e. in providing educational facilities for their children). Also included is the port-oriented transport system, geared to the colonial economy.

Seen in a larger context, this level concentrates on those parts of the *total* colonial space economy located in Britain and specializing in and dependent on the processing, manufacture, and shipment of colonially derived raw materials and also dependent on colonial markets. Both the colonial and metropolitan ports grow together: but in the metropolis, colonialism leads especially to the development of nearby port-related industrial areas exporting to the colonies.

This level also covers those areas benefitting from colonially derived investment. Whilst this issue requires much further investigation two examples can serve as an illustration.

The most obvious is the earlier reference to the so-called Lancashire 'cotton towns' and the Indian subcontinent. Here, particular localities processed raw cotton grown in India, Egypt (and earlier, the southern states of the USA), the export market for which was overwhelmingly in India and particularly, Bengal (now Bangladesh). Subsequently, the connection along this particular line in the spatial division of labour continues in the production and export of textile machinery for the Indian economy (Simmons and Kirk, 1981). Other examples in the early part of the twentieth century include West Yorkshire textile towns dependent on wool imports from Australia, New Zealand, and South Africa and equally dependent for exports on Canada and Australia; the dependence of the Midlands manufacturing industry (locomotives, machinery, etc.) on export markets in India, South Africa, and South America, or of ports, such as Glasgow and Liverpool on processing products (tobacco and sugar) from the Caribbean, or on shipping lines serving imperial routes.

The distinction to be made here is between regionally based industries and towns that operate in the full, open competition of world markets, and others (such as those mentioned) that not only operate within the politically and economically protected circumstances of colonialism but, by doing so, become locked into production forms aimed at the lower level of economic development that prevails in colonial societies.[2]

As the political ties dissolve with independence, economic growth occurs in the ex-colony at the expense of the particular locality in the metropolitan economy. Yet, as the colonial mode of production is replaced by harsher competition from other parts of the world-economy, structural, historical, and cultural ties continue: immigrant labour from low-income Asian countries is recruited to the same textile regions in order to lower production costs for a global market.

Performing a different function are those localities that benefit disproportionately from colonially derived investment. These may be particular towns and semi-rural areas (e.g. resorts, spas, 'the Home Counties'), favoured as residential locations by returned colonial expatriates. Whilst little research has been undertaken on this phenomenon (cf. King, 1976: 29), one example is significant in illustrating the continuity in the roles of particular localities in both the old and the new international division of labour. It demonstrates the colonial foundations of the internationally restructured contemporary urban system.

In the late-eighteenth and early-nineteenth century, 'new' Cheltenham was developed 'largely on the foundation of wealth acquired in the East' (Hart, 1981: 187). Money from East India Company operations in Bengal and elsewhere was invested in impressive residential property, urban development, and educational facilities in the town, which became an important 'basing point' for colonially derived capital as well as for the community of officials serving in different parts of the world. The development was to produce a lasting political flavour (as Cowan (1987) points out, it was here that Mrs Thatcher delivered her Falklands War victory speech)[3] and equally important, a spatially and architecturally impressive environment (the town centre marked by 'Imperial Square').

In the current phase of global restructuring, the economic, social, and, not least, environmental and architectural attributes of the town have been critical in determining Cheltenham's new role in the national and international economy. Benefitting from metropolitan decentralization, Cheltenham is now host to new producer-service industries, particularly in insurance, higher education (the Universities Central Council for Admissions, the Polytechnic Central Admissions Service, and defence (GCHQ)). It also contains the UK administrative headquarters for Gulf Oil. In addition to new investment in property, £9 million has been invested in conserving the early-nineteenth-century Regency environment (Cowan, 1987).

The urban level

This level of analysis concentrates on the economic, social, physical, and spatial development over time of the individual metropolitan city within the colonial system as modified by that system.

Here, the example may either be a regional city or the metropolitan capital. If the latter, the analysis will also incorporate phenomena listed at the international level (p. 60). In either case, we can hypothesize a city or town whose class structure (Foster, 1977) is related to its particular role in the old international division of labour, and particularly, the dependence of its working class on colonial markets (Massey, 1986a). Whichever way the social structure is related to the spatial structure (Dennis, 1984), from a viewpoint of the present day, one significant factor is that there were relatively few if any members of the peripheral, colonial population in the city (though ethnic diversity in the Jewish population existed in manufacturing centres such as Manchester or Leeds).

This is in contrast to the city in the colonized society (which received the products from the metropolitan) where there was a substantial, and largely bourgeois White 'ethnic minority' from the metropolitan society and an indigenous 'subproletariat' 'below' the working class in the metropolitan society. In other words, the international spatial division of labour located the underclass of the industrial city in the colonial society.

In the new international division of labour, analysis at this urban level focuses on the economic, social, and spatial restructuring of the individual city or town but takes into account its earlier history within a colonial mode of production. Here, a variety of phenomena need to be considered, some of which have recently been investigated (Cooke, 1986ab; 1989): the most obvious are international labour migration, especially from previously colonial territories (in the late 1980s, for example, 10 per cent of the population of Blackburn (in the traditional Lancashire textile region) is of South Asian origin, as is 25 per cent of the population of Leicester, previously an industrial/knitted-ware centre); new industrial structures include branch-plant manufacturing, decentralized white-collar employment, and office development, tourism, etc.

The urban-sector level

Analysis at this level draws attention to particular sectors in the city as affected by colonial or global capitalist processes: such sectors might be economic, social, or spatial. In the original formulation (King, 1976: 32), mention was made of those parts of the metropolitan capital, as well as parts in other cities, developed as a result of colonial connections. For example, in the early-nineteenth century, 80 per cent of the principal directors of the East India Company lived within three-quarters of a square mile of Central London (Marylebone, Baker St, Harley St); the large estates in South London (now public parks) of Sir John Tate (sugar interests), Edward Horniman (tea), and Sir John Anderson (P and O shipping lines) were all within the radius of a few miles; South Kensington was known to colonial ex-patriates as 'Asia Minor' on account of the large number of colonial-returned living there. St John's Wood was another favoured area (King, 1976: 32). Likewise, attention might be drawn to particular sectors of the urban economy with strong colonial links such as shipping, banking, insurance, real-estate services, or to the social and cultural influences of colonialism in the formation of elites, the production of science and culture (in museums), or the constitution of research and education.

Also included in this level are particular elements in the consumption sector–retailing, cultural consumption, and the specialized manufacture of luxury goods.

In the new international division of labour, this level of analysis provides wide scope for investigation. Focusing on spatial phenomena, attention can be given to spatial areas affected by different levels of international labour migration and continuities or breaks between colonial and capitalist world-economy processes.

Whilst these phenomena are discussed in Chapter 5, continuities can be seen between central areas (Kensington, Mayfair, and St John's Wood) benefitting from earlier investments of 'overseas capital' and contemporary investment from international elites (especially American, Arab, Asian, and European). Elsewhere, subproletariats previously located in the colonies are now relocated in previously working-class or lower-middle-class areas (such as Bengalis in the East End and West Indians in South London).

In industrial and commercial sectors, earlier colonially related industries (as in Docklands) are replaced by residential space for the international financial-services community or by branch plants of multinational manufacturers. In the financial sector, particular branches (banking, insurance, securities dealing) globalize markets that in many cases had colonial origins.

Outside the world city, analysis at sector level can also be undertaken in regional cities.

The unit level

Analysis of the influence of colonialism on urbanization and urban development at this level concentrates on individual institutions and the built forms to which they give rise. These may be economic, political, social, or cultural. Whichever the case, they would not be present in the urban system without the colonial connection.

In the imperial capital, these would include the institutions and buildings already referred to in the international (p. 60) level, for example, the School of Oriental and African Studies, the Colonial Office, the Commonwealth Institute, the Oriental Club, the School of Tropical Medicine, the Tate Gallery, the India Office (and the India Office Library), Canada House, Australia House — and innumerable others discussed in Chapter 5.

In the industrial towns or ports principally dependent on the colonial system there are sugar refineries, cotton mills, warehouses, and workshops as well as the housing for labour working in them, for managers and for the owners of the means of production. Other institutions and building forms required for social reproduction include schools, churches, chapels, trades union clubs, shops, etc.

All these buildings are originally constructed on 'greenfield sites', purpose-built for functional use in an industrial capitalist mode of production that, depending on sector and area, had closer or more distant relations to the colonial political economy: the factory specializing in sugar-cane machinery in Glasgow, sugar refineries in East London, jute factories in Dundee, cotton mills in Lancashire; other institutions (and buildings) have commercial functions (the Manchester Cotton Exchange); others have social and residential functions (the various Colonial Clubs or the universal bungalow; King, 1984a; 1986a).

In other cases, the connection is manifest less in the building's function or form but more in its name and architecture: 'Plantation House' (the London headquarters of a coffee and tea multinational, and now of a multinational conglomerate), 'India Mill' (a cotton mill in a

Lancashire town), 'Africa House' (a London office building for concerns with African interests).

In rural areas are the country houses and estates built on the profits of colonial enterprise. In the eighteenth century, these are the great houses of the 'Nabobs' such as Sezincote (Gloucestershire), Clive's house at Walcot (Shropshire), and some fifty others (King: 1984a: 66). Similar are the country seats of West Indian plantation owners such as Fonthill (Wiltshire) and Harewood (Yorkshire). In the nineteenth and twentieth centuries, another phase of country-house building continues, the finance often buttressed by overseas and colonial markets (Girouard, 1971; Aslet, 1982; see also King, 1989b, Chapter 7, note 2).

With the demise of the old colonial political economy, such buildings are either demolished or recycled for new uses. In the new international division of labour, the role of individual urban places, their labour markets, and potential for investment determines the nature of unit-level developments. In the world city, global control and management functions predicate office development. The investment functions of global cities mean that such buildings have not only use value as office space, but exchange value in terms of visual and symbolic content. They therefore attract a high degree of architectural input.

Reindustrialization through high technology relocates industrial employment away from old industrial areas or converts older industrial premises. Capital circulation and investment functions predicate consumption-oriented building and urban forms (Tabb and Sawers, 1984) whether as space-consuming, single-family suburban housing in greenfield suburbs of deindustrialized areas (King, 1984a; 1986a), out-of-town shopping malls and hypermarkets and in-town atria and galleria. The new service-oriented economy brings upwardly mobile residents into inner cities to gentrify previously working-class homes (Williams and Smith, 1986). Other new building forms developed for the capital-rich service economy include leisure centres, shopping centres, stadia, and yacht marinas.

With the disappearance of industrial roles in the old international division of labour, industrial buildings are demolished or recycled to meet the needs of the new consumer and tourist-oriented economy: old mills are converted for museums, for retailing, or for multitenanted boutiques. In rural areas, country houses become both residential and investment opportunities for the new international financial services industry, the bankers, brokers, and bond dealers (Thrift, 1987b), basing points for American universities moving into Europe, international centres for new religious movements, corporate conference facilities, investment opportunities for oil-rich Arab sheikhs, time-share developments or, as cultural objects, tourist-oriented museums (Sezincote is the home of a London merchant banker; Robert Clive's country house collection of Indian curios becomes a tourist museum).

In areas of economic decline or sections of the world city ignored by growth trends, old inner-city working-class terraces are abandoned by the old working class and taken over by new, international migrant labour.

* * *

What is apparent from these observations is that, whilst there is massive change between the old and the new mode of production, there are also continuities in the localities where they take place. The identity of these localities and the built environment that helps to create it, are instrumental in explaining why such continuities persist.

Part Two

IMPERIAL CITY: WORLD CITY: COLONIAL CITY

Chapter Five

DEPENDENT METROPOLIS

*Physical and social aspects of
London's role in the world-economy*

INTRODUCTION

London is the political, moral, physical, intellectual, artistic, literary, commercial and social centre of the world . . . no other city possesses the wealth, importance and abounding population which distinguish it. To London, as the true centre of the world, come ships from every clime, bearing the productions of nature, the results of labour and the fruits of commerce. . . . Its merchants are princes; the resolves of its financiers make and unmake empires and influence the destiny of nations.

Routledge's Popular Guide to London and its Suburbs, London: Routledge, Warne & Routledge, 1862: 1

As this quotation suggests, London was already performing the role of 'world city' in a nascent world system of production over a century ago (see also Briggs, 1963: 328); it continues to do so today. Yet especially since the Second World War, developments in that system have combined with changes in the international political order and revolutionary developments in communications and transport technology to both modify that role and bring major changes to the city. From one-time imperial capital, London has steadily become a specialized finance and business centre and base for cultural production in an increasingly integrated new international division of labour. The rise of financial empire has been accompanied by a drastic decline in manufacturing employment. As part of an overall population decline, locally born people have moved out of the centre to be replaced by the rentiers and representatives of international capital, international labour (from former colonies), and international tourists.

71

Greater London: Inner and Outer Boroughs

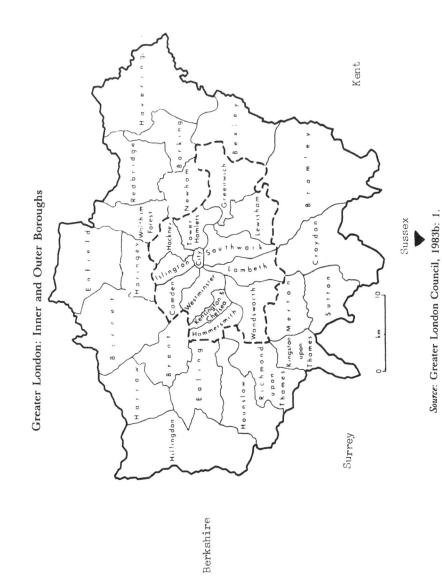

Source: Greater London Council, 1983b: 1.

Whereas eighty or ninety years ago, London was at the core of the world system, a generator of powerful economic, political, and cultural forces pushing out to the periphery, today it increasingly contends with equally powerful economic, political, and cultural forces pressing in from 'outside'.

It is a site for particular operations in the world-economy. It articulates the economic and political connections between the USA and the 'rich man's club' of the European Community. And because of its distinctive culture, privileged language, and past history, it retains its role as a post-imperial, international city, a centre for cultural production and ideological influence. In other respects, it increasingly takes on the characteristics of one of the old Treaty Ports of China.

The aim of this chapter is to set out some of these developments. Following a historical introduction, it identifies London's main function in the world system, particularly that of the City of London as provider of financial and business services. Subsequent sections examine sectoral employment change, the specialized role of the 'environmental professions' in the world system, and aspects of social restructuring. Throughout, particular attention is paid to transformations in the built environment consequent upon London's changing role in the world-economy.

LONDON, THE EMPIRE, AND THE WORLD-ECONOMY

By the end of the seventeenth century, London was the largest city in Europe, its economic foundations based largely on trade rather than industry. Estimates suggest that a quarter of its population depended on port employment in 1700; the growth of its trading wealth enabled the city itself to develop as a centre of consumption and to dominate English society (Wrigley, 1978). Routledge's comments (p. 71) probably anticipated rather than reflected the reality: in the first half of the nineteenth century, British trade, of which London was both the hub and the result, was predominantly with Europe and North America; but in the second half, with the massive extension of world communications through railways, steamships, and telegraph the structure of a more genuine world-economy was established with the Empire playing a prominent part (Woodruff, 1979).

The most obvious manifestation of this trade with an emerging world system was the construction of London's docks between 1799 and 1828. Their size, the range of commodities handled, and the origins and destinations of their shipping come through every London guide in the nineteenth century. West India Docks (opened 1802) covered 295 acres and was capable of taking 600 vessels (250–300 tons); East India docks (opened 1806) covered 30 acres, and could hold 28 East Indiamen and 60 smaller ships. London docks (opened 1805) covered 90 acres, cost £4 million to build, and took 500 ships. St Katherine's docks (opened 1828) covered 24 acres. Collectively, the docks accommodated over 1,400 merchant vessels and formed a major centre of employment (Mogg, 1841; Bohn, 1854).

The banking, financing, and shipping offices of the City of London were a direct outcome of this trade. The centre of the world's commercial gravity gradually moved from Venice, Florence, and Genoa to Bruges and Antwerp and before the end of the eighteenth century, London and Amsterdam were competing for the role of leading finance centre. From this period date the major institutions: Lloyd's insurance (1687–8), the Bank of England (1691), a stock exchange 'of sorts' (1670s), and the Royal Exchange (in the mid-sixteenth century) (Clarke, 1969; Ingham, 1984). These were the foundations of the capitalist City. By the mid-eighteenth century, colonial conquests in India, North America, and the Caribbean and a boom in the Atlantic economy made London the largest centre of international trade and its merchants the most prosperous in Europe (Anderson, 1987: 32, from which the following paragraph is taken).

The role of world financial centre came in 1815. The combination of the Industrial Revolution and the destruction, after Waterloo, of any competition to English hegemony brought into being a new form of world economy, needing, in Anderson's words, a 'central switchboard' to direct the flows of commercial exchange. For the first half of the nineteenth century, the changes wrought by the Industrial Revolution in the North were matched by the growth of the City in the South. From the mid-nineteenth century onwards, the City moved increasingly into overseas investment as its commercial and financial revenues grew, faster than the rate for the export of manufactures. Much of this investment went to the colonial empire (Anderson, 1987: 24).

With the investment went people. Altogether, between 1815 and 1925, over 25 million people emigrated from the British Isles, 15 million to the USA but 10 million to British possessions round the world (British North America/Canada — 5 million; Australia and New Zealand — 2.7 million; South Africa — c. 1 million; other areas —

1.2 million (Christopher, 1988: 37). Towards the end of the century in particular, these migration flows were urban to urban. They were also largely unilateral (though note also migration from Ireland to England).

London did, however, have a non-English population though despite its worldwide trade, this was still, in the 1880s, European rather than from other continents. As was still the case a century later, of an estimated 4.3 million inhabitants in 1887, the largest minority was Irish (81,000); the only other substantial groups were German (22,000), French (8,000), and Polish (Jewish) (7,000); Swiss, Dutch, and Italians numbered between 3–4,000 each; Belgians, Austrians, and Swedes, about 1 to 2,000; other nationalities were a few hundred or less (ABC, 1887). A scattering of Indian seamen and a small Chinese community lived in Limehouse ('Chinatown') and another small community of Afro-Caribbeans were in Canning Town (McAuley, 1987).

A quarter of a century later, London was seen to be much more cosmopolitan. In *The City of the World* (no date but about 1912) Edwin Pugh, with hardly hidden racism (and xenophobia) was writing:

> Foreign quarters abound, and living specimens of almost every race under the sun are to be met with in the streets in their native garb. There are Russians and Poles in their raw state in Church Lane off the Commercial Road; Orientals of every shade of complexion, from lemon-yellow to black, Turks, Moors, Kabiles, Armenians, Syranians, Persians, Hindus, Chinese, Japanese, Siamese, Malayans, Polynesians, and negroes, in Limehouse and Poplar; Italians at their Romish observances in Black Hill, Clerkenwell; Germans in Leman Street and Fitzroy Square; a whole cosmopolitan colony in Soho; and Colonials and Yankees everywhere.
>
> (Pugh, n.d.: 49)

IMPERIAL CITY

London . . takes a lot of understanding. It's a great place. Immense. The richest town in the world, the biggest port, the greatest manufacturing town, the Imperial city — the centre of civilisation, the heart of the world.

(H.G. Wells (1908) *Tono-Bungay*, London: Odhams Press, p. 73)

In the forty years before 1914, the growing importance of Empire as well as other changes in Britain's relations with an increasingly global economy were to make a noticeable impact on London's institutions. As the industrialization of competing nations accelerated, British attention turned to new, 'easier' export markets in Africa, Asia, and Latin America (Porter, 1979; Cain and Hopkins, 1980). Between 1831 and 1913, total British grain imports from continental Europe dropped from 86 to 5 per cent, by which time much of Britain's grain came from Canada, India, and the Argentine; 56 per cent of Britain's tea was from India. Exports also went wider afield, 40 per cent of them to Canada, the Argentine, South Africa, Australia, and New Zealand (Drummond, 1981; Woodruff, 1979). These connections were to be reflected in the capital.

In 1887, the building of the Imperial Institute of the United Kingdom, the Colonies, and India was begun on an extensive site in South Kensington. Its objectives were to exhibit the Empire's products and raw materials, to collect and disseminate information on trades, industries, and emigration, and to promote technical and commercial education in the context of Empire and further systematic colonization. The 280 ft tower, the conference hall, and other buildings were to provide the plant for the reorganized Imperial College of Science and Technology chartered in 1907 (Baedeker, 1889).

In addition to the goods flowing in and out of London's docks, imperial connections at the close of the century were also in terms of the import and export of both capital and labour. The emergence of a more mature international division of labour, transforming parts of peasant agriculture in Asia, Africa, and Latin America into monoculture plantation economies was a development of the four decades before the First World War (Hobsbawm, 1975; Wolf, 1982). It meant the movement, not only of capital but of bankers, engineers, colonial officials, army personnel, and planters around the world, the predecessors of today's international business elite. With industrialization and the growth of overseas trade, private and merchant banks in the city expanded to finance foreign trade and provide long-term capital for overseas development, a process that peaked in the half century before 1914 (Clarke, 1969; Christopher, 1988). By then, a capital surplus of some £4,000 million was invested abroad, financing railways and mines in Latin America, plantations in Asia, and urban development in Australia (Drummond, 1981; Edelstein, 1981). Almost half of this amount was invested in Empire.

With the capital went managers. Knowledge of the precursors of today's professional expatriate is scanty, yet, unique to London as capital of the Empire, were institutions providing both temporary accommodation as well as opportunitities for the exchange of information on the management of colonial economies, whether exploitative (West Africa) or settler (Canada, Australia). The Royal Colonial Institute (today, the Royal Commonwealth Society) was established close to government departments in Whitehall in 1868 to provide 'a place of meeting for all gentlemen connected with the colonies and British India'. Between 1880 and 1910, of some 110 leading clubs, some, such as the East India Club ('officers and Indian Civil Service'), the Empire ('Colonies and India'), the Naval and Military, the Oriental, the United Services ('majors, commandants, and above'), St James's (diplomatic service), Travellers', the Imperial Colonial, the United Empire, the Overseas Club, all within half a mile of Whitehall, catered for the administrators and proconsuls of the Empire (Baedeker, 1889; Ward, Lock, 1909).

At a different level, 'colonial agencies' articulated the supply of labour between the British provinces and the Empire overseas. Between 1881 and 1910, some 8.5 million Britons emigrated to Canada, Australia, New Zealand, and South Africa, as well as to Latin America and the USA (Barrat Brown, 1978). Monitoring the supply and providing information for the labour that flowed abroad, the agencies were located close to the main rail terminals at Victoria, Charing Cross, Blackfriars, and in the Strand (which still retains a strong colonial flavour). (In Victoria, separate agencies existed for the Commonwealth of Australia, New Zealand, South Africa, Canada, Tasmania, and Western Australia; in the Strand, for New Brunswick, Ontario, Queensland, Rhodesia, and Victoria. Not far away, were those for British Columbia, the Crown Colonies, New South Wales, Nova Scotia, and Quebec (Ward, Lock, 1909).

The exact relationship between building and urban development in London and capital flows within the Empire and the larger world-economy needs to be charted, as does their effect on London's employment. Garside (1984) suggests that after 1900, exports to the colonies increased as did returns on colonial investments, bringing a number of far-reaching effects. The revived export trades again focused attention on the docks where criticism of the facilities led to the establishment of the Port of London Authority (1909) with its large new building, including premises for Lloyd's Shipping and Register

Offices; 'serving and administering the Empire generated further employment in the capital, heightening the quality and diversity of the labour market' (ibid). In addition to 20,000 colonial administrators, numerous clerks, and professional advisers, 'countless others were engaged indirectly at home directing and supplying them mostly from London. The rising income from invisible exports — insurance, banking, and the rest — and from colonial investments was also channelled through London' (ibid.). In 1914, Britain was supplying 96 per cent of foreign investment in Oceania, 50 per cent in Asia, and 42 per cent in Latin America (Woodruff, 1979). In the years when capital investment overseas was relatively low (1890s–1905) building activity boomed; when foreign investment dominated (1906–10), there was a slump.

This investment function, and the City's specialized financial role in the world-economy had affected urban development for some time, leading to the building of specialized environments in specific places. By 1832, the City had already taken on much of its present form with the discount houses in place and several of the 'classical' fixed markets (Thrift, 1987b). In that year, Nathan Rothschild had observed that Britain 'in general is the Bank for the whole world . . . all transactions in India, in China, in Russia and in the whole world are guided here and settled through this country' (cited in Ingham, 1984: 93).

These specialized functions had a substantial effect in the first half of the nineteenth century. Between 1811 and 1861, the 17,000 smaller houses in the City were replaced by 13,000 larger ones, with the collective rental quadrupling (to £1.5 million) over the same period. In 1861, there were 619 blocks of buildings let out as offices and counting houses. Some of these, such as Gresham House, East India Avenue, and Mincing Lane Chambers, contained 'many hundreds of persons during the active hours of the day'. According to a House of Commons Report (1861), 'the last thirty years (1831–61) have seen the City of London nearly re-constructed by means of public works and private enterprise; and another twenty years will witness the completion of this transformation' (Scott, 1867: 49).

Between 1782 and 1861, the City had spent £7 million on public works and building. The entire frontage from London Bridge to Finsbury Pavement, intersecting the City from north to south, had been reconstructed, as also had the line from King William Street to the west along Cannon Street to St Paul's. Bartholomew Lane, Lothbury, and Threadneedle Street had also been rebuilt. In other places (Old Broad Street, Leadenhall Street) 'private enterprise and public companies have sufficiently contributed to the reconstruction of the City' (Scott, 1867: 50). In 1861, according to the Census, some 29,000 people worked there.[1]

Merchants	5,879
Bankers	263
Stock and commercial brokers	2,435
Ship-owners, brokers and agents	1,000
Accountants	1,910
Commercial clerks	17,225
	28,712

(Scott, 1867: 28)

In the 1880s, the Stock Exchange, with some 900 brokers and jobbers, was extensively enlarged (Baedeker, 1889); in the early 1900s, a rapid growth in shipping and insurance offices in the City occurred as both domestic and overseas insurance expanded. In Whitehall in the 1870s, the expansion of empire resulted in impressive new buildings for the Foreign Office, the Colonial Office (topped by imperial symbols: Morris, 1978), and the India Office. Increased expenditure on the navy to protect Britain's overseas investments and interests was behind the reconstructed Admiralty buildings in Whitehall (1895) and the rebuilding of the War Office (1898-1906) (Service, 1978).

The profits of colonial plantations in sugar, tea, or in shipping were to finance significant cultural additions to the capital. Henry Tate's sugar empire was behind the construction of the Tate Gallery (1895-7), H.J. Horniman's tea plantations behind the collection of ethnographic artefacts in the fashionable 'Arts and Crafts' building of the Horniman Museum (1896-1901) situated in South London's Forest Hill. While the connection between fluctuations in the imperial economy, the generation of employment and suburban expansion has not been precisely investigated, there are leads: Britain's pre-eminent position as 'first industrial nation' ensured that London remained the world's largest city throughout the nineteenth century (Weber, 1899); from just over 1 million in 1801, and 2.6 million in 1851 it more than doubled in size to some 6.5 million in 1901 (OPCS, 1973).

Extensive suburban development in West London in the later

nineteenth century followed specific building booms (Jahn, 1983) related to the relative attractiveness of overseas investment. The boom in trade, commerce, shipping, insurance, and foreign investment, and a twofold growth in central and local government staff between 1891 and 1911, all inflated the size of the white-collar sector for whom London's 'semi-detached suburbs' expanded (Jackson, 1973: 33). Inserted into the more ruralized, upmarket areas during these years was a new, specifically suburban form of consumer dwelling, directly inspired by colonial India — the bungalow (King, 1984a). In the early-twentieth century, industrial restructuring and economic developments on a national and global scale were behind massive suburban growth in London's south-east (Carr, 1982; King, 1984a).

Other evidence suggests contributions to suburbanization made by capital accumulated overseas: parts of London had particular colonial connections — Bayswater for colonial administrators, St John's Wood for the Indian Army, South Kensington for the colonial-returned (King, 1976). In the city, department stores — such as Whiteley's or the Army and Navy — benefitted from both colonial goods and colonial incomes. Burgeoning overseas trade combined with transport developments to fuel the habit of country living in what, in the twentieth century, was to become the 'stockbroker belt' round London in Surrey, Sussex, and Berkshire or even further afield. Profits accumulated by bankers, stockbrokers, overseas railway contractors, or the founder of the Home and Colonial Stores were to finance massive country houses, establishing in the process (with the help of colonial building abroad), the reputations of architects like Lutyens (Aslet, 1982) (see King, 1989b: 155).

Yet there is an obvious lag between London's relations to the world-economy and its representations on the ground. Before 1914, the 'new countries of settlement' and India accounted for 40 per cent of Britain's trade. After 1918, the significance of empire, both for imports and exports, increased (Drummond, 1981) and was manifest in redevelopments in the area around Trafalgar Square and the Strand: after Australia House (1914), came India House (1924), Canada House (1925), Africa House (1928), and South Africa House (1933).

The Empire continued to grow in size till 1937, its territories extended by conquests in the First World War. The British Empire Exhibition held in 1924 was both a celebration of British influence as well as a means to promote imperial trade. Its most significant contribution to London — Wembley Stadium — was the second giant building development devoted to spectacle as a regenerator of the economy (Harvey, 1987) (the first was the International Exhibition at Crystal Palace in 1851). Yet internationalization was now to take on a new dimension. The interwar years were to see a significant penetration from the USA.

THE (FIRST) EMPIRE STRIKES BACK

The arrival of American mass-consumption culture, as well as her retailing methods, is generally dated from the establishment of Selfridge's, on Oxford Street, by ex-Marshall Fields manager, Gordon Selfridge in 1909 (Weightman and Humphries, 1984: 25, from which much of the following is taken). It heralded increasing American influence in London, not only in consumption habits but in economic and industrial organization, life styles, and culture, not least through the dominance of Hollywood in the cinemas that sprang up at this time. Woolworths was to open in 1926.

The influx of Americans (and American wealth) made a significant impact on the West End, helping to displace an earlier social order with a new one. In the twenties, old aristocratic mansions in Park Lane such as Grosvenor House and Dorchester House were demolished, to be replaced by hotels and service apartments built on the New York model. Much of Park Lane and Grosvenor Square were rebuilt, the interiors of the new apartments particularly oriented to the growing American market.

On the industrial front, British firms, faced by increasing competition from American and German rivals, made economies of scale; consolidated, or effected mergers. In the early years of the century, the American Tobacco Company had attempted to buy up the British industry; this resulted, first in British manufacturers consolidating into the Imperial Tobacco Company and then, a merger between the two giants in British American Tobacco, whose new headquarters were set up at Westminster House on Millbank in 1915.

But a more substantial phase of restructuring got underway in the 1920s leading to more centralized control. The consequence was the emergence of enormous office blocks in which worked a growing phalanx of commuting managers and a large corporate clerical workforce. The result was to be seen in the north side of the Thames Embankment in Imperial Chemical House (headquarters of ICI) 'almost rivalling its neighbour, the House of Commons' (Weightman and Humphries, 1984: 38), Shell-Mex House, and Unilever House.

Another American, Irving T. Bush, conceived Bush House (completed at the south end of Kingsway in 1935) as a trade centre. One of the early tenants was the advertising agency of J. Walter Thompson. However, the trade centre did not take off and the BBC, whose own corporate symbol of Broadcasting House had been built in 1922, took over Bush House for its External Services Division, displacing the advertising role of JWT. Here, the 'Empire Service' was produced and broadcast until 1939 when, with some premonition, the BBC changed the name to the 'Overseas Service' and subsequently (1962) to the present 'World Service'.

As the old coal-based economy declined in the rest of the United Kingdom, the electricity-fuelled engine of London boomed, with two-thirds of all new jobs in Britain between 1923 and 1939 being created in Greater London. The drive behind this organization was international competition. Tariffs and the breakdown of free trade forced international competitors (principally American) to locate in the UK. And as the mass production of goods required a large market, they naturally set up near London.

With its population reaching its peak of some 8 million in the mid-1930s, London was the single largest consumer market in Britain, increasingly influenced by modern production, marketing and advertising methods, again, largely introduced from the USA. Of the 70 or so major US-owned factories set up in Britain between 1930 and 1939, over half were on London's Great West Road and North Circular Road, and on the Slough Trading Estate (Murray, 1985). These included Gillette's, Hoover, and Macleans. Others, like Ford, or Remington Typewriters (their 1930s portable boldly marked in gold lettering: 'Assembled by British labour in the Remington Factory, London'), were to the east. These were all strands of an increasing Americanization of the London economy and popular culture between the wars (Weightman and Humphries, 1984: 25–59); as late-nineeenth-century London had grown because of free trade so, in the 1930s, her economy grew because of the breakdowns in free trade (Murray, 1985). In the 1980s, the older scenario is being repeated.

FROM IMPERIAL CITY TO INTERNATIONAL METROPOLIS: THE INTERNATIONALIZATION OF LONDON. 1950s–1980s

Since the 1950s, and especially from the early 1960s, London's position in the international economy has radically changed. One insight into its changing function, and structure, can be gained by looking at its telephone directories.

The graph in Figure 5.1 charts the frequency of seven terms (International, Imperial, Empire, European, Euro-, World, Global) used as the first distinguishing adjective in the names of various organizations listed in the London directories (including corporations, public institutions, businesses, cinemas, clubs, restaurants, etc.). The graph shows the frequency of use at 10-year intervals between 1920 and 1987.

At one level of analysis, these descriptions can be treated simply as names that indicate, on the part of those who use them, a consciousness about a term, the meaning it is intended to convey and its perceived image and attraction. Looked at in this phenomenological sense (and recognizing that the data are determined by the availability of telephones), the table gives some insight into the changing cultural and geopolitical consciousness of London.[2]

In 1920, there were, coincidentally, as many entries under 'Imperial' as 'International' (108) and consciousness of 'Empire' (41) was twice as great as of 'European' (12), 'World' (10), and 'Global' (0) combined. The use of 'Empire' and 'International' increased more or less in tandem until 1940.

Ten years later, however (1950), 'International' entries (201) were already twice those of 'Imperial', the latter going into steady decline from the 1940s. Also at mid-century, 'global' makes a first and notable appearance with a single entry (Global Tours); and the prefix Euro-, non-existent in 1920, has five entries. By 1960, 'International' entries were way ahead and 'World' listings (73) were about to overtake both 'Imperial' (78) and 'Empire' (68), each of which was soon outnumbered by 'European' in the mid-1960s.

In the quarter century between 1960 and 1985, 'International' listings more than doubled (from 211 to some 440). 'Imperial' and 'Empire', however, continued to decline in use as, from the mid-1960s, 'World', 'European', and especially from the 1980s, 'Global' names relegated both 'Imperial' and 'Empire' to the bottom of the list. Whilst

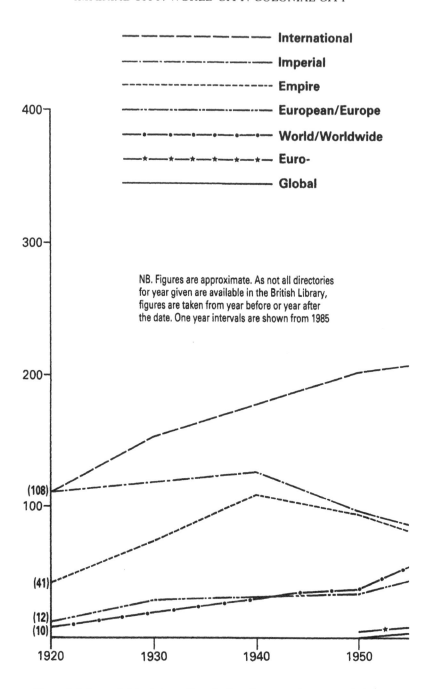

Figure 5.1 Imperial to international city: the consciousness of change

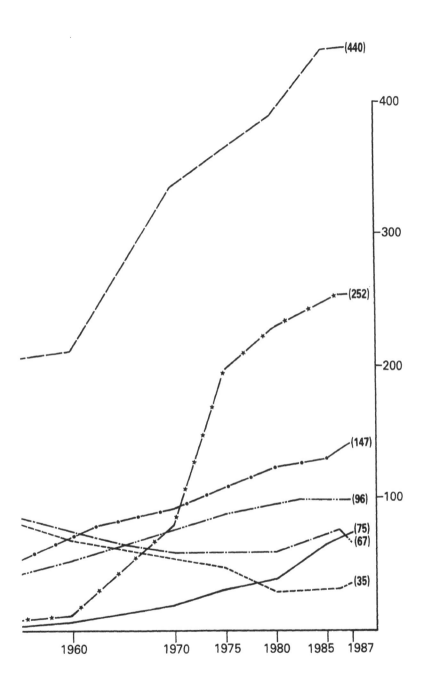

'Imperial' has recovered in the last few years, at 67 entries in 1987, this is only 15 per cent of the 'International' listings.

Other things happened between 1960 and 1970. 'Imperial' entries were outnumbered by both 'Japan/Japanese' (a mere 10 in 1960) and 'American'. (The significance of this is referred to subsequently.)

	1960	1970	1987
Imperial	76	60	67
Japanese	10	64	60
American	62	71	87

In 1987 neither 'France/French' (12) nor 'German' (40) figure as prominently.

In a different, and more sophisticated form of analysis, descriptors such as 'International', 'European', and 'Global' provide a useful, if crude index of the growth of 'genuinely' international or European organizations whose international operations are organized from London.[3] These can be looked at in more detail.

Of some 440 'International' names in 1987, at least two-thirds can be referred to as genuinely international operations and between 1960 and 1987, the number of such organizations grew by some 300 per cent.

Whilst requiring more detailed and sophisticated research, what these data suggest is that as the number of 'International' organizations in London steadily increased, those listed as 'British' declined, from a peak of 898 in 1970 to 637 in 1987. This prompts at least three speculations: either 'British' institutions have changed their name (even becoming 'International'), or have completely disappeared, or have moved out of the capital. In the case of many 'national' and public-sector organizations, this latter explanation is correct as various non-profit organizations, particularly charities, have been driven from London by rising costs, replaced by private, foreign, or 'international' concerns. (At the time of writing, the Economic and Social Research Council was a recent example.)

These phenomenological data need to be understood in the context of massive changes that have taken place not only in London's but also in the UK's changing position in the world-economy since 1950. (The following draws on the brief but useful account by Hudson and Williams, 1986: 2-12; see also Marder and Alderson, 1982: 245.)

In the early 1950s, the UK was still committed to maintaining its old international political and economic role, sufficiently preoccupied with the imperial benefits of the old international division of labour to reject membership of the European Economic Community. About half of all Britain's exports went to the Empire, whose captive markets took 75 per cent of her manufactured goods.

By the late 1950s, however, and increasingly into the 1960s, with competition increasing in the newly independent countries of the old Empire — from the USA, West Germany, France, Italy, the Netherlands, and Japan — the UK's share of world trade in manufactured goods dramatically declined (it had been about 33 per cent in 1900 and was 9 per cent in 1955). As a response, major UK-based multinationals had already begun to invest heavily overseas, with industrial capital following financial capital in its traditional international orientation. By 1960, 30 major UK-based multinationals had subsidiaries abroad.

Whilst trade with non-EEC countries grew in the 1960s, the decision to enter the Common Market (later, the European Community) in 1973 totally restructured the earlier trading patterns of the UK and, in turn, was instrumental in changing London's role, both in relation to the European space economy as well as to the world-economy as a whole. As imports from 'developing countries' (the old Empire) declined from 53 per cent (1955) to 9 per cent in 1982, imports (of manufactured goods) from the Community rose from 15 to 51 per cent of the whole in the same period. The UK economy has been integrated with that of the Community at the relative expense of links with the USA and the ex-colonial Commonwealth countries, with increasing disparity between UK imports of manufactured goods from the European Community and exports to it. In 1983, for the first time, the UK recorded a deficit in the trade balance in manufactured goods. By then, her share in world trade in manufactures was 5 per cent. Tourism has become a major source of income (see p. 124) and the City (whose share of the world market in 'invisibles' is also declining; Thrift, 1987b) has also moved to regain lost ground.

It is in the context of these larger structural changes that the present role of London in the world-economy — with its associated effects — can be viewed.

THE CITY IN THE WORLD-ECONOMY

The primary function of London in the world system today is that of the City of London as banking and finance centre. With deregulation in 1986 and the full internationalization of securities dealing, this function (including the operation of markets in foreign exchange, Eurocurrency, insurance, gold, shipping, and commodities) has become much more pronounced. Closely allied are the production and export of services in accountancy, law, and business, and a general clearing-house function for financial information to and from the rest of the world. Once largely confined to the 'square mile' round the Bank of England, in the last three years (1984–7) these functions have expanded massively to the west and, through Docklands, to the east.

In 1987, the obvious symbol of this role was still the City's (and the country's) tallest building, the 600 foot Natwest Tower housing the international subsidiary of Britain's second largest bank. The continued global expansion of branches, the construction of a new Natwest headquarters building in New York (1982), and the opening of a division in Miami all reflect the global integration of capital. In 1982, almost half of Natwest's outstanding loans of £28,000 million were made overseas and income from international operations (£1 billion 1980–2) contributed a third of its profits (National Westminster Bank, 1982; 1983; CIS (Counter Information Services), 1983). Yet in 1985, Natwest's symbol (and indeed, the City as a whole) was threatened by the proposed vast, new, American-inspired development to the east, in Canary Wharf. The 10 million sq ft, £3,000 million worth of office space consisted principally of three skyscrapers, half as tall again as the Natwest Tower and the tallest in Europe. (The scheme, the product of American developer G. Ware Travelstead and architects I.M. Pei, Kohn Petersen Fox and Skidmore, Owings and Merrill, is referred to on pp. 99, 152).

Though London's history as an international capital centre dates mainly from the early nineteenth century, between 1960 and 1980, the international market came to dominate City activity, especially in banking. The City has replaced its dependence on sterling business by securing access to much larger international sources of capital. It has done so 'by welcoming subsidiaries of major financial institutions from other countries and by establishing a wholesale market in their international deposits' (Jones, Lang, Wootton, 1980). These developments have immensely accelerated since deregulation in 1986.

The effective internationalization occurred in two ways; on the one hand, the UK domestic economy with which the city is interdependent, is itself dependent on the world-economy; second, with the City's specialized function as provider of services to the world of international finance, 'more than any country in the world, its fortunes are strongly determined by the growth and activities of other nations' (Jones, Lang, Wootton, 1980). Hence, levels of growth in world trade are seen as the best predictors of demand for, or decline in, the City's services: these in turn are significant explanations for change in the stock, as well as rents, of City office space. For an insight into the transnationalization of capital and the particular role of London in this, the rapid growth of foreign banks since the 1960s is worth examining in some detail.

The internationalization of London banking

London's nineteenth-century importance as a centre of trade was a natural attraction to foreign banks. However, it was imperial expansion that was responsible for some of the first — from Hong Kong (James Capel, 1808), Australia (Australia and New Zealand, 1835; Westpac, 1864, and National Australia, 1864), New Zealand (the Bank of New Zealand, 1862 and the National Bank of New Zealand, 1872), again from Hong Kong (the Hongkong and Shanghai, 1865 and the British Bank of Middle East, 1865), Canada (the Canadian Imperial, 1867 and the Bank of Montreal, 1870). From the 1860s, other countries came in: the Ottoman Bank (1863), Comptoir d'Escompte de Paris (1867), the Crédit Lyonnaise (1870), Société Générale (1871), the Swiss Bank Corporation (1898), and the major German, Italian, Belgian, and Russian banks. In an earlier spurt of internationalization around the turn of the century, these were followed by American banks: Chase Manhattan (1887), Morgan Garanty (1892), and Citibank (1902), with the Bank of Japan arriving in 1898 (*The Banker*, November, 1987).

By 1914, there were about thirty foreign banks in the square mile round the Bank of England (which employed 900 staff in the 1880s compared to some 5,500 a century later, Clarke, 1969; Baedeker, 1889; *The Times*, 19 July, 1983).

As with other London banks, their interests were in financing trade, funding long-term capital, and providing services previously undertaken by British banks: some were representative offices. By the 1930s, there were over 80 and by 1961, over 100.

Since then, the growth has been phenomenal. Numbers doubled between 1961 and 1971, and again, between 1971 and 1981; in 1987, there were 453 either directly or indirectly represented, a number that has decreased slightly in the last three or four years (*The Banker*, November, 1987). These ranged from the Citibank, with 3,500 or more staff, to two-man offices, though the majority have from 50 to 400 employees.

However, with 'Big Bang' in October 1986 there was a massive increase in staff employed by foreign banks, practically doubling in numbers from the 38,000 in 1983 to 72,000 in 1987. This figure (which also includes securities houses) represented a 36 per cent increase over 1986 and was a direct result of deregulation and trading in securities. After the Stock Market crash of October 1987, there was an equally drastic reduction. The other consequences of this are discussed in the Conclusion.

Whether traditional banking houses, merchant-bank subsidiaries, or large consortia, the number of banks has expanded as part of the general globalization of capital, concentrating banking and bank-related services in the three 'supranational' key cities at the core. With the rapid growth of the Eurodollar market since the 1960s, and especially, the mid-1970s, the main functions of branches of foreign banks such as trade, finance, and serving subsidiaries of home-corporation customers, has been overtaken by entrepôt finance. As with other corporate institutions, banking has diversified, constantly widening its services, including international merchant banking with medium-term loans to companies on a global scale.

As Anderson (1987: 68) states, these developments have been a direct outcome of government's policy towards the external linkages of Britain to the world-economy. The abolition of exchange controls in 1979, emancipating finance from its remaining national constraints, had dramatic effects on the City. In the 1970s, the Eurodollar market had grown through the very mechanism that had initially triggered the industrial downturn in the manufacturing areas of the country, the OPEC price rise, the bulk of whose earnings were subsequently deposited with London banks. It was the huge quantity of the financial flows involved in recycling petrodollars through the City that proved such a magnet for foreign (especially United States) banks to come to London as a base for operations.

But in addition to the Euromarkets, commodity markets and foreign-exchange dealing have attracted foreign banks to London. But it is the Euromarkets that, in Goodhart and Grant's view (1986: 9), have halted London's decline as a financial centre and returned it to its nineteenth-century status as the capital of international finance.

These markets (in which financial instruments such as Eurobonds, Euro-equities, and Eurocommercial paper are bought and sold with Eurocurrencies, that is, any currency held outside its country of origin, such as a dollar held in a London bank account or Eurodollar) are big: in 1986, the Eurobond market raised 200 billion dollars of capital compared with 6 billion dollars of new shares raised on the Stock Exchange in 1985.

The Eurodollar deposits have also attracted big American banks to London because London is freer of regulation than New York. And because American (like other foreign) banks do not have to deposit a proportion of their assets (at zero interest rates with the Bank of England as is the case in most domestic banking systems), it was cheaper for foreign banks to make their international loans from London than elsewhere. Hence, foreign banks distribute Eurobonds all over the world to raise money for governments and companies. The Eurobond business is overwhelmingly in the hands of the foreign banks in London.

The growth of foreign banks in London has also accompanied the establishment of multinationals. As *The Banker* put it 15 years ago (November, 1973): 'As the number of businesses which operate at a truly multinational scale proliferates, so will the need for institutions able to meet all their financial requirements world wide'.

The increase in the number of banks results from the global mobility of capital through international communications as well as the growth of specialist services. London's historic role as the hub of the world's financial system has resulted from a number of advantages: the accessibility of support services such as foreign-exchange brokerage, expertise in financing international trade, and good communications. Also, British practice does not divide investment banking from commercial banking, as American and Japanese laws do; London is the one place that speaks English (as Zurich, Frankfurt, and Paris do not) within a time zone that fits in well with Tokyo and New York in a 24-hour trading clock (*Economist*, 11 October, 1986). Earlier there was the traditional attraction of compacting institutions within the 'square mile' of the City, but with new technology, this has been extensively stretched in the last five years. Equally important has been the Thatcher government's commitment to *laissez-faire* and the lower levels of regulation than in other centres.

At a time of increasing competition from other world cities, these are seen as its major advantages. As a result, in the early 1980s,

London topped New York, Singapore, Zurich, and Hong Kong, both in numbers of international banks and the volume of dollars, yen, and pounds that pass through it annually (CIS, 1983: 26). In addition, however, British entry into the EEC and activities in the Eurocurrency and Eurobond markets have been behind the proliferation of foreign, especially continental and American banks. North Sea oil, overseas-project finance, and the growth of public-sector borrowing have also contributed. American banks have moved into the British property market with loans for mortgages and construction and into other areas, for example, the steady expansion of time-sharing since 1975.

Loans to developing countries also contributed — until the system started to crack. From 1973, the volume of British bank loans to the Third World increased at a rate of 20 per cent per annum; though British bankers were less committed to South America than those of the USA, British loans amounted to £11.3 billion; one of Britain's four largest banks had loans of £900 million to Mexico, a sum equivalent to 60 per cent of shareholders' funds (CIS, 1983: 28). This was in the 1970s when lending was 'North to South'. After 1982 and the debt crisis of developing countries, together with the huge growth of the Japanese surplus, lending has moved from 'East to West', from Japan to the USA and Europe. Capital has moved from institutional savers to 'creditworthy' borrowers and major banks, in both the USA and Britain, have written off millions of Third-World debt. Now, Japanese and other capital floods into the UK — with major effects on urban development.

Since 1971, there has also been a rapid growth in 'ethnic banks' (particularly Asian) partly connected with the growth in international labour migration. Though initially catering for immigrant communities in London and the industrial and deindustrializing Midlands and the North, these have later, because of cultural and religious links to the Middle East, made inroads into domestic lending. From 1973-4, the oil crisis and the boom in oil money led to the establishment of Middle Eastern banks (branches of these now occupy some of the most prestigious locations in London, at the Hyde Park corner end of Park Lane, in houses that, before the First World War, were owned by the aristocracy). Large American banks have moved into the oil industry and have serviced UK subsidiaries of American manufacturing companies.

Between 1975 and 1982, the total assets of foreign banks as a proportion of total assets of all banks in the UK had risen from 52 per cent to 61 per cent (*The Banker*, 1982). Till the early 1980s the major influx of banks was from the USA but between 1983 and 1987, the number of American banks decreased from seventy-six to fifty-nine whilst Japan, with numbers growing from thirty-three to forty-five in the same period, made a major impact on the City. By mid-1987, the Japanese share of overseas non-sterling lending from London had risen to 41 per cent and their 25.5 per cent share of all banking assets booked in London was more than even the London clearing banks' 24.8 per cent (*The Banker*, November 1987).

But whilst London is the world's largest international banking centre, its share of total foreign lending by banks has fallen from 29 per cent to 24 per cent between 1975–85. In those years, business has shifted to other time zones, especially in offshore centres.

In short, there is fierce competition in both banking and securities trading in the City, principally from the Japanese and the Americans. According to Goodhart and Grant (1986: 9) there are fears that the UK financial services industries could go the way of the motorbike and motor industries.

As in other spheres, London has become the arena of international capital, the site for the creation of global profit.

The internationalization of securities dealing and 'Big Bang'

Despite this build-up in international banking, commodity, Eurobond, and foreign-currency markets since 1960, the historic and carefully protected monopoly exercised by the Stock Exchange over securities dealing, with its fixed commissions and monopolistic control, was to distinguish London as a world financial centre from Chicago and New York.

Since the 1970s, securities dealing had become increasingly transnationalized. The abolition of fixed commissions on the New York Stock Exchange in 1975 had led to increasing international competition and led to their progressive abolition in London. In the 1970s especially, foreign-securities houses were allowed to set up in London and later, outsiders could buy a 29.9 per cent interest in broking firms (*The Banker*, 1983). Yet in terms of turnover, London's share of global securities dealing was one-fifth that of Tokyo's or one-twelfth that of New York (Thrift, 1987b). The London market was losing out to competition.

The immediate occasion for deregulation and 'Big Bang' was

essentially a political decision by a government committed to free-market principles and is well known (for an account, see Thrift, 1987b). In a deal with the Conservative government, the Stock Exchange agreed to abandon fixed commissions on securities dealing and to open up its doors to overseas members rather than face legislative charges under Restrictive Practices legislation. Hence, whilst the two central features of 'Big Bang' were the abandonment of fixed commissions and the adoption of 'dual capacity' (eliminating the distinction between brokers and jobbers), it also heralded a wave of associated changes that totally transformed the way the City operated, the actors partici-pating, their organizations and, not least, the accommodation required; and with millions of pounds spent on electronic hardware, the equipment used.

The most important of these has been the expansion of large financial conglomerates from the US, Japan, and elsewhere (some anticipating the changes somewhat earlier) combining banking, dealing, and currency trading as well as other financial functions and buying up the expertise of the Stock Exchange firms to become the new 'market makers' of 1986. The number of securities houses in the City (a mere 10 in 1960) which, with the development of international trading, had tripled in the 1970s, increased to 121 in 1987 — especially in the last two years. Where previously only the Americans were predominant, the Japanese now control 37 out of the 121 securities houses (see Tables 5.1 and 5.2).

The largest of these conglomerates employed substantial numbers of personnel including the many dealers: the Americans, Merrill Lynch (1,400), Shearson Lehman (1,370), Salomon Brothers (900), Goldman Sachs (700), and the Japanese Nomura International (600) and Daiwa (400). The effects of the crash of 18 October were to lower these — and numbers in other houses substantially in late 1987. The large American and Japanese conglomerates, however, were much larger (in terms of stock-market capitalization) than were the locals: Nomura (at around 30 billion dollars), Salomon (6 billion dollars) compared with 1 billion dollars each for the British Warburg, Kleinwort Benson, or Hambro (*Economist*, 11 October, 1986).

The influence of the local (thatcherite) political climate has been crucial to these developments. Since the abolition of exchange controls by the Conservative government in 1979, the City's income from investments abroad, reflecting massive outflows of capital, has greatly expanded. (Though it should be noted that all the foreign investment

Table 5.1 Foreign securities houses in London, 1960-87

1960	10
1970	27
1980	76
1983	94
1987	121

Source: The Banker, November 1983 and November 1987.

Table 5.2 Foreign securities houses by country, 1978-88

	1978	1983	1988
United States	31	48	37
Japan	9	23	36
Canada	12	14	11
Australia	3	11	12
Switzerland	1	2	5
France	0	2	4
Others	1	7	16
Total	57	107	122

Source: The Banker, November 1978, 1983, 1988.

that flowed out was handled by foreign brokers (Goodhart and Grant, 1986: 7). The drop in sterling also increased the value of earnings in foreign currencies. Between 1981 and 1983, the City doubled its surplus on overseas business (from £2,300 million in 1980), principally by net overseas earnings and insurance companies. Banks' income from foreign investments doubled each year between 1979 and 1983. Likewise, income to insurance companies from investments abroad increased over 20 per cent between 1979 and 1983. Pension funds invested abroad have also trebled: the value of total assets held by UK pension funds increased fourteenfold in fourteen years, from £10.6 billion in 1971 to £150 billion in 1985 (*The Times*, 8 October 1986). Banking, insurance, pension funds, and commodity trading account for some 80 per cent of Britain's net invisible earnings by the private sector.

In November 1987, the Bank of England *Quarterly Bulletin* indicated that Britain's overseas assets (many arising from the investment of North Sea oil revenue), at £114.4 billion, were greater than those of any other country, including Japan. It is these that have refuelled City redevelopment at the expense of investment in the domestic economy in the regions.

BANKING, TRADING, AND THE PHYSICAL-SPATIAL TRANSFORMATION OF THE CITY

These developments have had a major impact on building and the land market not only in the City but also in London as a whole, in the South-East, and, through decentralization of office development, to places of over 200 miles around. For the present, however, we can focus on the City.

Between 1965 and 1980, some 13 million sq ft of new floor space was developed, about half on new building, half on replacing old; the traditional banking area around the Bank of England was extensively expanded (Jones, Lang, Wootton, 1980). Banks and financial institutions took 85 per cent of the largest prime office developments completed in the 1970s, a concentration that contrasted with a far wider spread of non-bank, non-financial users in the major City developments of the 1960s (Barras, 1981). In the twenty years to 1982, more than one hundred office developments of over 100,000 sq ft were completed or planned (*Chartered Surveyor*, June, 1982), and between 1980 and 1985, a further 8.6 million sq ft were built (Lemon, 1987).

Even before 'Big Bang', City space had become increasingly specialized, suggesting the rapidly increasing dependence of the City of London on its continued role in the capitalist world-economy. Before the widespread adoption of information technology, the traditional banking and finance area of the 'square mile' centred on the Bank of England, with the main demand for space being within a radius of 300 yards. Whilst the need for face-to-face transaction in this area provided a logic for location, prestige was obviously of equal importance. The language of realty agents referred to 'prime banking areas' (e.g. Threadneedle Street), 'acceptable banking areas' (e.g. the south end of Moorgate), and 'locations not historically acceptable to the banking fraternity' (Jones, Lang, Wootton, 1980). In the late 1970s, the growing number of banks with increasing space requirements (1,000–2,000 sq ft for representative banks, 5,000–10,000 sq ft for those of branch status, 100,000 sq ft for a major office) either led to higher rents or cracks appearing in the traditional system of location.

On the one hand, social and cultural factors encouraged location on specific streets, putting constraints on lateral expansion; on the other, height restrictions and the preservation of statutorily defined 'historic' structures or ones deemed by a planning elite as of 'architectural interest' inhibited vertical growth. Hence, there was increasing conflict of interest between the representatives of international capital to demolish buildings and redevelop at higher densities or greater heights, and what were seen as national, or local interests, defined by local planners (Dyer, 1982) more interested in preservation. Lateral growth had stretched the boundaries of 'the City' with larger, more prestigious banks moving to lower-rent areas on the fringe, out of the City to the West End or, in certain cases, out of London completely.

An early example of this 'pre-Big Bang' development, manifesting the investment of foreign capital and cracking open the traditional boundaries of banking's 'square mile', was the announcement of 'London Bridge City' in October 1983. The £400 million project, 'the largest single commercial development in Western Europe', was undertaken by the property subsidiary of the Kuwait government's investment bank at Hay's Wharf, on previously derelict dockland, despite public protests about excess development and the need for industrial jobs and housing (*The Times*, 17, 18 October, 1983). Also breaking previous conventions concerning expansion on the south bank of the Thames has been the 12-acre development at Butler's Wharf (Lemon, 1987).

Making space for the 'Big Bang'

The deal between the Stock Exchange and the Tory government was sealed by the Conservative election victory of 1983 (see McRae and Cairncross, 1985). The effects were soon apparent: mergers and disposals among banks and broking firms started to occur. And as some firms began to see the spatial implications, *The Times* property correspondent wrote that the City property market was 'showing signs of an upturn' (27 September 1984). Yet even early in 1984 the City of London planning authority continued to take a strong line against further redevelopment and, despite opposition from both developers and financial interests, maintained a strong conservationist line in favouring 'historic buildings'. City office rents had remained fairly stable from 1973 to 1983, with the demand for space by foreign banks being steady at 20 per cent per year.

Within eighteen months, this stance was totally reversed. Fearing that London, strangled by controls, would lose out to other world cities (especially Tokyo and New York) the City planning department made sweeping changes. Moreover, by 1985, the revolutionary proposal by American developer G. Ware Travelstead, of 10 million additional square feet in the London Docklands Development Corporation scheme at Canary Wharf posed a major threat to City rents.

Early in 1986, the City planners conceded that 'plans for controlling the "square mile" over the next decade were conceived in ignorance. They had not appreciated the changes the financial revolution would bring' (*Financial Times*, 21 February 1986). Moreover, much of the property world had been just as ignorant, though the ignorance had begun to fade by the end of 1984. In what might be seen as a desperate attempt to (literally) gain some lost ground, four drastic modifications were proposed by the City planners to increase city space by 25 per cent and add a further 17 million sq ft.

These were to include decking over London Wall and Upper Thames Street, along the northern and southern fringes of the City and clearing some post-war blocks; various road decking and bridging-over proposals including buildings by the London Museum, Holborn Viaduct, including the topping of Cannon Street Station with offices and a heliport: these and other sites were seen to be able to provide thirty City locations for the new 'banking factories' of upwards of 250,000 sq ft that the major conglomerates — with their massive dealing rooms — required; third, as 20 per cent of City floor space was already underground, to make further, and deeper, excavations below ground level; and finally, to dispense with a whole catalogue of listed 'historic buildings' previously thought worthy of protection (*Financial Times*, 4 July 1986).

Such proposals, however, were irrelevant for some of the larger American conglomerates. The delicate combination of prestige, spatial requirements, and cost that hedged the carefully discussed decisions governing location on the fringes of the City in the 1970s, have been abandoned by banks, which now have many different buildings. According to the *Financial Times*:

> American conglomerates moving into London have little interest in rabbit warren Victorian blocks or older glass towers but rather, want layers of dealing floors buried in wide efficient structures. Hence, the move of Salomon Brothers to 200,000 square feet over Victoria Station and the interest of Morgan Stanley in the new office towers in Canary Wharf.
>
> (*Financial Times*, 4 July 1986)

The effects of 'Big Bang' have had a major impact on the City, on what used to be called the 'City fringes', as well as the West End, the central region and, by knock-on effect, the Greater London property market. In late 1986, property specialists Edward Erdman were predicting that demand from the conglomerates could lead to the redevelopment of one-third of the 140 million sq ft of space in the 'square mile', the City of Westminster, and their surroundings in the next five or ten years. By the end of 1987, 11 million sq ft of new offices would have been added to the City and its fringes in the context of the Big Bang (this, without taking into account the proposed 10 million sq ft more at Canary Wharf). The extent of the change can be measured by the fact that in the three years between 1985 and 1987, permissions were granted for *five times* the amount of office floor space in the City than in the previous three years, 1982–4.

Whilst all this occurred before the market collapse of October 1987, the frenzied speculations are comprehensible. In November, while 'not a sod had been turned at Canary Wharf', the development was still deemed to go ahead, but subsequently, was taken over by New York's largest landlord, Toronto-based developer, Olympia and York. Though still at 10 million sq ft, the design had been changed, the skyscraping towers lowered. It was to offer space at 50 per cent of City rents and plenty of it was already taken up (*The Banker*, November, 1987). Subsequent developments are discussed in the Conclusion.

Big Bang buildings

The arrival of these huge financial conglomerates in the last two or three years, and the transformation of the City into a global capitalist casino, has had a major effect on both the buildings and the fabric of the City. To guarantee the City's place as a global office centre, the City of London plan relaxed controls over new building. Plot ratios (the amount of floor space in a building expressed as a multiple of its site area) were raised to a limit of 5:1 on the City fringes. But it is not just the amount of space that is required but also its scale and quality. As Duffy (1987) writes, where previously City institutions would give rise to ten 50,000 sq ft buildings that would fit into the existing street pattern, the new 'banking factories' require 500,000 sq ft in one slab. Big financial multinationals such as Citicorp, American Express, or Nomura, using London as a global base, need such buildings in order to function.

The introduction of information technology from the late 1970s has been central to these developments. Yet, where earlier speculation suggested that this would lead to the dispersal of office functions, it has paradoxically, according to Duffy (1987), brought the City together: the new buildings are part of the computer. Major changes have occurred in buildings compared to ten years ago. The requirement of huge trading floors (the largest for 600 desks at Salomon Brothers) has meant that floors of 50,000 sq ft have replaced those of 5,000 sq ft. The pre-Big Bang mergers have created larger organizations and these have consumed more space, with organizations often moving on from one building to another. New technology (cabling, services, additional requirements to ensure heat loss, and accounting for 40 per cent of the cost of the building) have also taken up more space. In the mid-1980s, the requirements of National Westminster Bank were put at 10 million sq ft.

Equally important in the appearance of the new mega-buildings have been different sources of financing them. Previously, building size was often limited by the street plan, the planners' decisions, and the need to break investments into manageable and affordable units. Now, with global capital seeking profitable outlets for investment (see p. 92) and new financial instruments for funding buildings (such as unitization), these old constraints have been eroded. The size, nature, and scale of buildings is decided by the requirements of international capital.

'GLOBAL CONTROL CAPABILITY': LONDON'S MULTINATIONALS

The growth in multinational banks has been accompanied by a growth in multinational headquarters, as also branch and regional offices, not only in London but in the larger region that it dominates. These developments reflect the gradual incorporation of London (and the UK) into an increasingly integrated European and world division of labour (Murray, 1985). With the goal of achieving the European Community's 'internal market' set for 1992, these trends have intensified.

Both for domestic as well as foreign multinationals, London's locational advantage is threefold: it gives immediate access to domestic markets; by its geographical proximity, it gives access to those in Europe, and historically, its traditional links with the Commonwealth make both it and Britain 'an ideal location from which to serve a wide range of markets overseas' (Dunning and Norman, 1987: 625). According to recent research, the 'existence of a large potential market' and the 'need for personal presence' are the two most important factors governing the location of international regional and branch offices in the UK (ibid).

Morgan (1961) in that year concluded that 'the control of large scale British industry had its principal focus specifically in the West End of London', though significantly at that date he made no reference to multinational holdings. Since then the function of London as a centre of multinational control has greatly increased both through domestic mergers and the inflow of foreign capital and institutions. Though the movement of major company headquarters out of London receives considerable publicity, the trend in favour of concentration is encouraged by the growth of multiplant firms through diversification and takeovers, as well as multinational location policy (see also, Hoare, 1975). Of the 51 British companies listed among 430 of the world's largest multinationals (Dunning and Pearce, 1981) about 80 per cent had their headquarters office in central London or, for a handful, in its immediate vicinity. The concentration is predominantly in the West End, especially in Victoria and Westminster, with the other concentration in the City.[4]

In addition to multinational banking and insurance, other producer services have arrived or extended their operations in London in recent years. US firms have sought skilled labour, particularly top graduate labour in the natural, physical, engineering, and social sciences in Britain, and also *information*, especially financial information, which particular locations, such as the City, supply. Hence, US firms invest in management consultancy activities to poach labour for parent companies and also in investment banking, stockbroking, foreign-exchange sectors, to gather information (Dunning and Norman, 1987). (According to the *Times Higher Education Supplement*, 18 September 1987, only 40 per cent of Cambridge engineering graduates go into industry; the rest are persuaded into the City, the civil service, and associated spheres. In 1987, 10 per cent of UK graduates were entering accounting.)

Research by Dunning and Norman (1983) suggests the corporate logic and competitive advantage behind the establishment of branch and regional offices in London. Data on the distribution of offices of US-based business-service activities, for example, shows 73 branch banking offices in central London compared to Paris (32), and Brussels

(18); in management consultancy and executive search, 40 offices in central London compared with Brussels (39) and Paris (19); in legal practice, 24 in central London, compared with 10 in Paris.

Regional offices 'acting as a sounding board for parent headquarters and providing managerial expertise for the European operation' (Dunning and Norman, 1987) were located in London because of its good communications, availability of professional staff, suitable office accommodation, and nearness to European branch offices; access to a major political and business centre and an international airport were also necessary. Operating costs played a secondary role: political stability and supporting services were seen as more important.

More recent research by Dunning and Norman (1987) shows that the main factors influencing the locational choice were, in order of importance, airport links, language, social and cultural factors, the availability and cost of accommodation, and the 'business framework'. Where English is the global business language, this is an important influence, not only for the US and the Japanese, but also for European and Scandinavian companies. Though levels of personal and corporate taxation were important, sympathetic attitudes to foreign corporations by government were more so. The Conservative government's 1979 relaxation of exchange controls 'had removed an important disadvantage of UK location'.

The attraction of London for multinational headquarters has continued despite large discrepancies in the costs of office space compared to alternative locations in Europe. (London's office rental values have long been the highest in the world.) In 1987, they were 30 per cent higher than in New York, about 40 per cent over Paris and 5 times those in Brussels (Hillier Parker, 1987). One explanation is the perception of London's importance in the world of international finance; another, however, is that labour costs in offices have been lower than in other European cities. The UK is perceived as a 'low-wage but high-skill economy' (Dunning and Norman, 1987).

One study, in the early 1980s, showed salary costs for secretarial, clerical, and managerial grades in London at between a half and one-third lower than those in possible alternative locations in European cities; for senior executives, they were much lower, at between one-third and a quarter those of European levels (Barras, 1981). Combining labour and rental costs, Central London locations were still considerably lower than those of alternative locations. Though cost factors, as indicated here, are not seen as the prime locational determinant, the effect of the continued internationalization of office space is to distribute capital in favour of office developers at the expense of employees' wages and salaries.

The example of one area, Victoria Street, running between Victoria Station and the Houses of Parliament and close to government offices and other commuting stations of Charing Cross and Waterloo, demonstrates the impact of these developments on the physical restructuring of London. In the last 25 years, this has been virtually rebuilt, and principally relet on the profits of global corporations.

Originally constructed in the mid-nineteenth century, it became the headquarters area for firms of consulting engineers, professional associations, and building consultants whose location near the House of Commons at a time of rapid railway development was significant in terms of lobbying requirements on railway legislation (Langton *et al.*, 1981). Accommodation was in small chambers of some 2,000 sq ft in five- or six-storey mansion blocks. Over the last twenty-five years, redevelopments have created new office space for leading world companies. In the 1980s, these included the UK subsidiary of the USA's Monsanto Chemicals; the major oil companies (BP House, Esso House, Mobil House, as well as accommodation for Arabian Gulf Oil and National Iranian Oil). Other major clients included British American Tobacco, American Express, major banks and property companies (ibid). Subsequently, BP and Esso were to move out of London.

Whilst there has been a continuing decentralization of middle- and lower-grade office employment from Central London to the suburbs, the rest of South-East England and to surrounding regions (Barras, 1981), especially East Anglia and the south-west (Hall, 1987), according to Lemon (1987) the scale of this decentralization has been exaggerated. Nevertheless, at the end of 1988, it was reported that a growing number of firms were moving their headquarters out of Central London. Of those with more than 100 employees, half of such moves were to other parts of Greater London and eight out of nine were to somewhere in the south-east. The result was 'a bigger and richer south', the beneficiaries, a 'crescent of small towns from East Anglia to the southwest, about 70 miles out of London' and an hour's train journey from the centre (from Peterborough round to Swindon) (*Economist*, 17 December, 1988, p. 27).

Even with these moves, however, this concentration of corporations is a major determinant in the polarization of occupations and life styles, both between 'north' and 'south' as well as within the region. It also has significant effects on other aspects of the city.

FOREIGN ASSETS: THE INTERNATIONALIZATION OF LONDON PROPERTY

In the last fifteen years, and especially from the 1980s, the financial function of London's buildings (storing capital rather than people) has been increasingly transnationalized: banks, hotels, retail centres, apartment blocks — have become receptacles for both the national and international surplus. According to a report of the Investment Property Databank (1986), 71 per cent (by value) of all property investments by pension funds and insurance funds are in the south-east of England and nearly half of the total 'institutional investment' buildings are in London (*The Times*, 17 August 1986).

The forces behind this internationalization of investment are both general and specific: the internationalization of capital generally, the expansion of international producer services in London, and the internationalization of property markets (Daniels, 1986; Thrift, 1986a). The growth of multinational enterprises, the international spread of pension funds, the search for safe havens for capital fleeing from Hong Kong or South Africa are accompanied by particular instances: the investment of oil surpluses after the oil-price rises of 1973, or ten years later, the shift by American, Japanese, and British banks from Third-World lending to domestic lending.

Whilst these international investment shifts were already taking place in the 1970s, little notice of them seems to have been taken until well into the 1980s (the now-defunct Greater London Enterprise Board, for example, had no information on Arab investment in London in 1984 (personal communication, Robin Murray)). Yet, following the oil-price rises of 1973, Arab buyers had bought clubs, restaurants, houses, and shares in property companies and had, in 1976, an estimated 20,000 flats in London and were spending some £200 million a year on property (*The Times*, 15 January 1985).

In the early 1970s, the Kuwait Investment Office (set up in the 1950s) was investing an estimated £100 million a year in property and company shares. Whilst it is not known how much investment the KIO handles, the *Observer* believed it to be 'very large, if not the largest' comparing it to the Prudential Insurance combine's of £17 billion. In 1982, it was estimated that Kuwait had 42 billion dollars invested in foreign equities, including £528 million in British companies. The General Manager of KIO regarded London as 'a major international investment centre' (*Observer*, 27 January 1985).

In 1974, the Kuwait royal family acquired St Martin's property company for £107 million (equivalent to one week's oil revenues) through which subsequently, major property acquisitions and development were undertaken. The most spectacular of these, London Bridge City, was announced in Autumn 1983, a £400 million scheme for the redevelopment of half a mile between London Bridge and Tower Bridge on the south bank of the Thames to create about 800,000 sq ft of office space (a large part of it later taken up by the American Citibank). In 1985, this was described as 'the biggest commercial development in the capital since the Great Fire of 1666'. The scheme also included a private hospital, retail space, and houses (*The Times*, 15 January 1985), pushing out local interests in favour of industry and housing in the process. St Martin's also bought the Piccadilly Hotel.

Other spectacular buys included the Saudi purchase of the eighteenth-century Crewe House in Curzon Street, once the home of the Liberal Marquis of Crewe, for use as an embassy for £50 million. An attached 'motor house' for ten cars was subsequently built (*The Times*, 7 January 1987). In 1977, Beechwood, a mansion set in 12 acres in Hampstead Lane, Highgate was bought by the late King Khaled of Saudi Arabia. In 1986, this was estimated to be worth £8 million (*The Times*, 22 September 1986); The Holme, the Regency mansion in Regent's Park, till 1984, occupied by the University of London's Bedford College, was sold by the Crown Agents to an Arab buyer (on condition it was used as a single-family dwelling) for £5 million. (Cuts in the public-education budget forced Bedford College into a shotgun marriage with Royal Holloway College and exile to a less expensive site near Staines, well outside London.)

In one, if not the most exclusive and secluded embassy compound rows in London, Kensington Palace Gardens, a house was sold to 'a Middle Eastern buyer' for £5 million in mid-1985; Witanhurst, in West Hill, Highgate, 'the largest house in London after Buckingham Palace' also went to an Arab buyer for £7 million in 1985 (*The Times*, 22 December 1985).

Other Middle-Eastern interests included the Vauxhall Cross site on the South Bank (scheduled for 1.5 million sq ft of offices), the Effra site fronting the Thames at Vauxhall (again, for 1 million sq ft of offices) and innumerable schemes in Docklands.

Later reports ('Arab buyers learn caution', *Financial Times*, 28 August 1987) indicate that Middle-Eastern investment in UK residential property is fading though many such houses are for use as much as investment. However, there are cultural (and also sociopolitical) factors that have determined which properties are favoured by Arab buyers. These are likely to be in outer rather than inner London with a closed perimeter wall giving security as well as privacy. The ideal target for a married man from the Gulf states, according to the author, is a modern building with 6–8 bedrooms either in London or within a three-quarter mile drive (and within striking distance) of Heathrow. Architecturally, these should have a large room to use as *majlis* or *diwan* where guests can be received apart from the family's private rooms, a requirement easier to achieve in horizontally laid-out houses, not tall narrow ones with stairs, of which Arab buyers are not fond. Hence, these requirements (together with a fear of inner-London crime) favour the better-class suburbs.

In the 1970s, Arabs had bought country houses as investments, sometimes paying highly for them. When times got worse, parts of the estate were sold off for redevelopment. According to the *Financial Times* (28 January 1987) 'it has not mattered if their view is spoiled. Arabs are more concerned with the indoors than outdoors'. In this way, however, for English buyers, the property, without the estate and 'the view', had lost its potential value.

With tourism a major growth industry, as well as London's function as international conference and business centre, the hotel industry is another realm for international investment. Whilst tourism has steadily increased, the promotion of London as a conference venue has, with over 300 conferences in 1986, placed it second only to Paris in the international conference stakes (two major conference centres have been completed in recent years; in 1986, the London Conference Centre hosted some 18,000 delegates of the American Law Society).

The 1987 Hanson Trust purchase of the American Intercontinental chain, with its 3,000 hotels in 40 different countries, is evidence of the hotel as a favoured investment area, whether for the US, the UK, Japan, or Australia, a practice that has significant effects not least in terms of employment structure. But the internationalization of hotels also has equally important effects on wages (see p. 124), social and cultural practice, consumption styles, and knock-on effects in many interlinked industries. Between 1981 and 1985, 60 of London's leading hotels changed hands; though only some of these involved external capital, they were the biggest and most well known.

The Dorchester Hotel, sold in 1976 for £9 million by its British owners, McAlpines, was bought and sold four times over (by Lebanese, Saudi, other Middle-Eastern and American interests) before being purchased by the Sultan of Brunei for £43 million in 1985 (*The Times*, 17 January 1985). Shortly after, it was bought by the American group, Regent International Hotels (managers of fourteen hotels in Asia, Australia, the Pacific and the USA) for £45 million. In 1984, Marriott Hotels (with 147 hotels around the world) bought Europa Hotel, as part of their expansion plans in the UK and also bid for the Piccadilly (*The Times*, 29 May 1986). The American Holiday Inns, the Hilton and the Sheraton, also expanding in Britain, accounted for thirty openings in 1984. In 1986, the Los Angeles-based Trafalgar Holdings bought the International Hotel in London. In June 1985, the American Ramada Group was reported to have set aside £100 million to spend on British hotels with London as a main target; other chains — including the Holiday Inn and Sheraton, Marriott — were also in hot pursuit.

A prime instance of London's changing fortunes being determined by forces beyond its control is the case of the Millbank office of the Crown Agents in the heart of Westminster. The Crown Agents, a 300-year-old institution, were responsible on the part of the state for the purchase and supply of equipment and materials to the Empire (and its successors).

In 1983, Brunei declared its independence from the Commonwealth. In July of that year, the Sultan dismissed the Crown Agents from their role as manager of Brunei's £3000 million investment portfolio of oil and gas wealth, establishing in their place an independent Brunei Investment Agency, advised by two American banks. The decision led to the loss of over 300 of the Crown Agents' civil-service jobs and a move from their central Westminster headquarters in Millbank (*The Times*, 2 August 1983; 29 October 1983). In 1985, the building was bought by the American Raleigh Enterprises for £10 million to be turned into a hotel (in 1987, discussions were taking place to privatize the Crown Agents).

Linked to the takeover of hotels is the purchase of prestige stores. Of these, the most significant was the buying of Harrod's by the Egyptian Al-Fayed brothers for £313 million in 1985. The Al-Fayeds (one of whom has lived in the UK for twenty years) are also owners of the Paris Ritz, the Rockefeller Plaza in New York, as well as property in London's Park Lane and Paris's Champs Elysées. In Bond Street, still a mecca for international tourists and world centre for 'Old Masters' and jewellery, some of the main fashion stores were bought by Danish oil magnate, Peter Bertelson. According to the manager of Gucci, 60 per cent of the trade is foreign. Though forty of the stores belong to Prudential Insurance, as customers, over the last two or three decades, the English aristocracy have been replaced by Americans and customers from the Middle East. A branch of Tiffany's (New York) opened in 1986 (*The Times*, 12 March 1985).

In August 1985, the Mitsubishi Estate Company, the real-estate arm of Mitsubishi, the Japanese financial conglomerate, bought its first commercial property in Britain, paying £34 million for a bank occupied by them in King Street, in the heart of the City. The building was sold by the property subsidiary of Jardine Matheson. The deal was arranged by Jones Lang Wootton, the first British estate agents to set up an office in Tokyo. In August 1985, *The Times* property correspondent indicated that 'Japanese investment in British property is rare' (*The Times*, 9 August 1985).

Elsewhere, international capital rebuilds prominent stores. Whiteley's in Bayswater, once the mecca of colonials at home and the world's largest department store, was in 1986 being transformed by a consortium of British and Far-Eastern (i.e. Singaporean, Hong Kong, and Malaysian) developers in a £25 million, 253,000 sq ft retail project (*The Times*, 2 June 1986).

The developments in the City, outlined on pp. 96–100, pushed up rents by 50 per cent in 1986–7 and created, in the words of The *Banker* (November, 1987: 53) 'a galloping bull market in City of London property'. Most of the money for these developments has come predominantly from Japanese banks but also, from Scandinavian, French, and some 'second flight' US banks. In many of the really large deals, however, a large percentage has been Japanese: the Broadgate development was 40 per cent underwritten by Japanese lenders and between 1985 and 1987, Japanese banks' investment in UK property had increased eightfold to around £250 million. 'At least six major Japanese banks were actively chasing property business' (ibid).

RESIDENTIAL PROPERTY

Overseas investment in domestic property was already advanced long before the influx of American, Japanese, Australian, and other bankers came in with the 'Big Bang'. Advertisements in *The Times* for 'American bank urgently requires 1- to 4-bedroom properties in Belgravia, Chelsea and Knightsbridge' go back well into the early 1970s, though they have increased rapidly in the 1980s. In 1984, *Saville's Magazine* was indicating that the percentage of 'the best property in Central London' (a somewhat indeterminate category) going to overseas buyers had increased from 40 to 60 per cent between 1980 and 1984 (in 1984, 26 per cent of buyers were from Europe, 19 per cent from the USA, 12 per cent from the Middle and Far East and 3 per cent from Africa). Chesterton Prudential's *In Residence* (9 May 1987) indicated a somewhat lower figure of some 30 per cent of 'prime residential properties' in Central London going to international buyers between 1980 and 1985. In 1980, the main buyers were from Europe (25 per cent), the Middle East (22 per cent), the USA (16 per cent), and the Far East and Africa (each 14 per cent). In 1985, Americans had been prominent especially for apartments in the £100,000–£200,000 range because of the weakness of the pound in that year. Hong Kong buyers were more interested in transferring money from the colony into property in the £200,000–£400,000 range. The pattern for international bankers was to stay between two and five years, buy a country home, and send the children to school in Britain (*Saville's Magazine*, Autumn, 1984).

By 1984, this level of investment had already made houses costing £1 million common with three 'Millionaire's Rows' being referred to in London — the Bishop's Avenue, Hampstead, Kensington Palace Gardens, and Avenue Road, St John's Wood. As more overseas buyers came in, so residential property was increasingly designed and marketed for the international market. In *Property International* (1985: 1, 2) (begun in 1985 and financed by Middle-Eastern money), Million Dollar Homes Ltd of Curzon Street advertise 'The Georgians', a new 'white house in the prime part of Bishops Avenue' ('white' refers to the colour of the facade. The design, with huge classical portico, would fit easily alongside low-country South Carolina plantation houses.).

Three years later, as the 1992 fully integrated Common Market drew nigh, other continental developers were marketing Dutch-style homes, 'subtly evoking the atmosphere of Amsterdam' in a development named Vermeer Court, Rembrandt Close in the remodelled docklands (*The Times* (Property Section), 29 October, 1988, p. 9).

Beverly House, a new apartment block in Park Road, opposite Regents Park offering 'spacious flats and a high level of security' at prices (mid-1986) of up to half a million pounds, were designed by American-based architects, CRS. Apartments and houses in prestige locations 'well presented, with a high standard of finish and attention to security' are bought as long-term capital investment. Americans are said to account for 80 per cent of short-term lettings in this prestige section of the London rental market. All flats had video entry phones, underground garage, and security grill and access by lift to the entrance hall (long-term rents were £600 per week for three-bath, three-bedroom units (*The Times*, 10 September 1986)). Increasingly, apartment blocks are being built specifically aimed at the overseas or British investor, both as long-term investment with the option of high-rental income for company lets.

In 1985, half a million pounds was given as the average price of 'a top-quality London house in a prime residential district' (a figure long since past). Between 1981 and 1985, Arab buyers were seen to be the main purchasers of these London houses, but, according to *Property International* (1985), overseas buyers fell into two categories, though in each case, perhaps using the property only for a few weeks in the year as one of a chain of homes around the world.

New purchasers wanted to be as central as possible, within half a mile of Hyde Park, Mayfair, Kensington, Knightsbridge, or Bayswater. On the other hand, those of a few years' standing were tending to move out to the leafier suburbs. The demand for prestigious central London locations has caused Westminster city-planners to turn down planning applications for change of use from residential to commercial. 'A wave of refurbishment is under way as owners . . . upgrade their properties to modern luxury standards attracted by prices of up to £1 million' (ibid, 1 February 1985, p. 31).

In 1985, five new flats marketed by Chesterton's with views across Grosvenor Square were sold to a Greek buyer for £800,000 (for a twelve-room apartment): an Indian bought two flats for £1.5 million. The other buyers were Iranian and American.

Internationally oriented property deals have been promoted by the rapid growth in the 1980s of a new range of giveaway real-estate house journals, glossy brochures, expensive advertising, and marketing hype in the new range of 'country house' magazines (King, 1987b). These developments (increasing business for the large printing and design industries in London) have been promoted by the estate agents whose commissions have risen to finance them. In an advertisement feature (in the *International Herald Tribune*, 25 October 1985) promoting

Central London agents, it was reported that in anticipation of 'Big Bang', prime London residential property had increased by 35 per cent in 1984–5. The customers of international banks, the multi-nationals and the conglomerates, were establishing their headquarters and homes for senior staff in London, 'the natural English-speaking base for North Americans to penetrate the polyglot Common Market'. Many were said to be seeking flats to accommodate and entertain clients and customers.

Other factors, however, were seen to be important. 'The medical facilities offered by Harley Street, the elitist standards of independent schools, the ancient universities as well as the social calendar embracing Derby, Wimbledon, Henley and Ascot' which enjoy 'global status followed by the beau monde throughout the Season. . . . No wonder the tax exiles are returning and mixing with the new millionaires created by the Unlisted Securities Market competing to establish luxury homes' (ibid).

Different nationalities have settled in different areas. Kensington has been of special attraction to the Arab interest who, in 1985, were interested in houses in the £650,000 bracket in 'quiet areas' such as Phillimore Gardens. Homes were preferred with good security and space to accommodate servants in the basement or ground floor. Advertised in the same edition of *Property International* (1 September 1985) is a house in Princes Gate (SW7) offering 13,000 sq ft with five major bedroom suites, three self-contained staff suites, and including an infrared security system, electronic shutters, video entryphone, (heavy) window grilles (the Regency front looking like an adapted prison), and use of communal gardens, all for the price of £3 million.

Outside Central London was another 'highly specialist market fed by strong international demand' (ibid) in Wimbledon, Kingston upon Thames, Hampstead, Highgate (with substantial family homes with eight bedrooms and banqueting hall for seventy-five guests), and Regents Park.

As Thrift (1986b) has shown, major estate agents, already operating on a British, European, and Australian scale from the 1950s and 60s, greatly extended their operations in the next decade. Between 1972 and 1982, Jones Lang Wootton, for example, opened further European, Far-Eastern, and American offices in:

1972	Amsterdam	1975	New York
1973	Frankfurt	1978	Los Angeles
	Antwerp	1980	Houston
	Hong Kong	1981	Chicago
	Singapore	1982	San Francisco
	Kuala Lumpur		Washington DC
	Penang		

Likewise, estate agents Richard Ellis, who also had a strong presence in Africa from the late 1960s (Johannesburg, Cape Town, Durban, Pretoria, and Harare), also expanded throughout Europe, the Far East, and North and South America

Amsterdam	1973	Chicago	1976
Madrid	1974	Atlanta	1976
Hong Kong	1978	Dallas	1982
Singapore	1976	San Francisco	1980
Jakarta	1982	Rio de Janeiro	1979
New York	1982	São Paulo	1980

Whilst these operations were primarily concerned with commercial (especially office) developments, they — like other agents (Chesterton's, Saville's, etc.) — are typical of outlets marketing residential and commercial developments in London.

When the *Financial Services Act* was passed in 1986, mergers between financial institutions and estate agents, leading to the acquisition of chains of estate agents throughout the country, meant that the network of property covered for the international buyer moved from London and the south-east (and occasionally 'the country') to the entire national market. Local markets became national, and national ones, international (King, 1987a).

These developments, however, have only been part of the massive internationalization of property investment and ownership that has characterized the 1980s. British companies have become major investors in United States property, particularly in Washington: London and Leeds (a subsidiary of Ladbroke Group PLC) has commercial developments worth more than one billion dollars in Washington DC, New York, and Coral Gables, Florida. Olympia and York, the Canadian developers of Canary Wharf in London, were in 1988 New York City's biggest commercial landlords, with over 12 million sq ft of office and retail space. The Japanese construction firm, Kumagai Gumi, is partner for a 225 million dollar office tower in Seattle as well as the new 'World Wide Plaza' on the site of the old Madison Square Garden in downtown Manhattan. Because of the decline in construction in the oil-based economies of the Middle East, construction companies with an equity stake in what they build search for new projects worldwide (Wald, 1988). Holding foreign property, like bank assets, has become capital's conventional wisdom in the late 1980s, with 16 per cent of United States bank assets owned by foreign investors (Tolchin and Tolchin, 1988).

ECONOMIC RESTRUCTURING AND EMPLOYMENT

This increased specialization of London as the premier financial centre in the new international division of labour as well as centre for multinational corporate control has had a major impact on employment — and the lack of it. On the one hand, in the twenty years between 1968 and 1987, the number of staff employed by foreign banks and securities houses has increased sixfold, from 9,000 to 72,000 and City of London employment as a whole to 390,000 (1981) (*The Banker*, November 1986; November 1987); in London as a whole, over 20 per cent of the labour force worked in banking, insurance, and finance in 1988 compared to 15.9 per cent in 1981. In the south-east region outside London, a further 11 per cent are in the same sector compared to 7 per cent in 1981 (Massey, 1988). On the other hand, manufacturing, which employed almost 1.5 million people at the beginning of the 1960s, in 1985 offered jobs to less than 600,000 and by 1990 this figure was expected to fall to below half a million. According to Murray (1985), 'manufacturing is effectively finished for London and should not be saved'. In the mid-1980s, London had 400,000 registered unemployed, there were increasing numbers working in the sweated trades in East London and with the growth of homelessness, and the 'informal economy', comparisons with Third-World cities had become increasingly common (GLC, 1985; Mitter, 1986ab; Murray, 1985). On 15 December 1987, a report on London's homelessness in *The Guardian* was headed 'Beggars on the streets of London'. Like New York in 1988, London had the highest house prices and the highest levels of homelessness; among its employed population, it had the highest levels of inequality in earnings and the sharpest rate of increase in inequality (Massey, 1988: 15).

These developments can be looked at in more detail. The relentless shift to services, and decline in manufacturing, has taken place against a drastic fall in population in Greater London since its peak of 8.6 million in 1939. From just under 8 million inhabitants in 1961, the population had fallen to 6.7 million in 1981 though has risen slightly since. Decentralization to the zone immediately surrounding Greater London (the Outer Metropolitan Area) and to the South-East region had reached 10 million by 1981. Yet the most outer ring, and especially the zone immediately surrounding Greater London, are growing at a decreasing rate (Hamnett and Randolph, 1982).

Yet in the late 1980s, these figures have become increasingly academic. So great is the pull of London and the South-East in general, that its daily commuting reach now extends to a radius of well over 100 miles, to Newark (even Doncaster) in the north, Cardiff and Bristol in the west, and the major towns on the south-east coast — all well within two hours from London. Inter-city trains, which in the early 1970s took people to London for the occasional day, are now as full as the commuter tube: bags of tools can be seen as well as executive brief-cases.

Describing 'The London Effect', *The Sunday Times Magazine* (3 January 1988) referred to 'towns which until a year ago thought they were in the Midlands now find that they are part of the commuting south-east and traditionally dozy seaside resorts are trading bathchairs for BMWs'. Led by the London clearing banks, employees had been offered zonal allowances as far north as Bedford and Saffron Walden and as far west as the Hampshire–Dorset border, with the obvious effect on house prices which, between 1985 and 1988, increased by 25 per cent per annum. In 1987 in London, a standard 'Victorian semi' costing £250,000 was increasing at £137 a day.

Between 1985 and 1987, Grantham, 110 miles and 80 minutes by train from King's Cross station in London, changed from 'a rural market town to a commuter suburb', the 6.00 a.m. trains from Yorkshire (Leeds) 200 miles from London (also with occasional, rather than daily commuters) taking on the social (and spatial) character of commuter specials as they arrive there at 7.20 a.m. With electrification of British Rail's Eastern region cutting journey time to under two hours, Yorkshire is being pulled into the same vortex. Similar developments have transformed many south-coast resorts as well as East Anglia.

These developments have made major changes to the social and physical character of the regions affected, not simply in massively increasing house prices (20 to 25 per cent per annum in the last three years in many places), changing the form and size of dwellings constructed, and altering the social composition of towns, but have equally exacerbated problems of homelessness in semi-rural areas. They have also had a major impact on the use of the London Underground; as the disaster at King's Cross station brought to light in November 1987, numbers of passengers on the underground increased by 65 per cent between 1982 and 1987 (from 498 million to 769 million a year). This was followed just over a year later (December 1988) by another commuter disaster, with 33 people killed in a British Rail crash at Clapham Common. The huge increases in commuting, the peak demand passing record levels set in 1970, has been accompanied by shrinking levels of public subsidy to British Rail, seen by the Conservative government as a necessary strategy in its plans to privatize the State-run concern (*Economist*, 17 December, 1988, p. 29). The spatial spread of London's employment, therefore, is both politically controversial as well as difficult to define. The data that follow must, in this case, be read with some caution.

As Table 5.3 indicates, between 1961 and 1978, total employment in Greater London fell by 17 per cent; within this movement, however, manufacturing employment fell more drastically by 47 per cent. Since then, these trends have accelerated, with total employment down to 3.5 million in 1981. Whereas Britain lost 25 per cent of its manufacturing jobs in the decade between 1971 and 1981, the decline in London was much greater — 36 per cent overall and 41 per cent in Inner London (GLC, 1985). As a percentage of the total numbers employed in London, the number of manufacturing employees fell from about 34 per cent (1961) and subsequently below 20 per cent (Table 5.4) in 1981 (GLC, 1981b; MSC, 1982)

Combining all non-industrial jobs together, the services sector accounted for 81 per cent of London's employment in 1981, reflecting the main concentration of many of the country's service occupations there. As a percentage of total UK employment in particular sectors, London had 37 per cent of employment in banking, 37 per cent of 'other financial institutions', 38 per cent of 'other business services', 31 per cent of insurance, and 28 per cent of property-owning and management employment (GLC, 1981; MSC, 1982). Employment in financial, professional and miscellaneous services was the only employment category to increase (by 5 per cent to 1,468,000)

Table 5.3 Changes in manufacturing and total employment, 1961–1978

	1961	1966	1971	1974	1978	1961–78
Manufacturing employment		thousands				% change
Greater London	1429	1309	1093	940	769	– 47
Rest of SE Region	990	1142	1200	1194	1092	+ 10
England and Wales	7626	7848	7442	7248	6513	– 15
Total employment						
Greater London	4386	4430	4084	3990	3679	– 17
Rest of SE Region	3240	3775	3900	4149	3612	+ 12
England and Wales	20,913	22,325	21,562	22,186	20,186	– 4

Source: Young and Mills, 1983.

Table 5.4 Greater London occupation groups, 1961–1981
(Numbers of employed in thousands)

	1961 % of total employment		1971 % of total employment		1981 (est.) % of total employment	
All industries and services						
Operatives	1507	34.3	1164	28.4	844	22.8
Office workers	1404	31.9	1527	37.4	1606	43.4
Others	1475	33.8	1393	34.2	1251	33.8
Total	4386	100	4084	100	3701	100

Source: Young and Mills, 1983.

between 1973 and 1983 (GLC, 1985: 4).

Yet as Hall (1987) has pointed out, London's increase in financial and business services employment has been well below the national average, and much below that of other major cities. None the less, taking *inner* London as the nearest approximation in size to the large provincial cities, Hall's figures indicate about 78 per cent in services and 45 per cent of employment in the 'information sector'. With the exception of Edinburgh, this is 5 to 10 per cent more than in most other major British cities and reflects the growing importance in London of functions of planning, marketing, advertising, sales, etc. Manufacturing decline in London, and especially in the inner city is part of a larger metropolitan, national, and also international picture. Initial explanations for these changes (Damesick, 1980; Evans and Eversley, 1980; Golland, 1980; Gripaios, 1977; 1980; Knight *et al.*,

116

1977; Lomas, 1978) were sought in terms of industrial relocation in Britain but by the late 1970s it became clear that the substantive context for restructuring was the transnationalization of the economy.

According to Murray (1985), the main causes of industrial decline in London were not the redistribution of industry in Britain but in the shift of industrial production to the continent and beyond. In market terms, it represented the declining competitiveness of British industry and reorganization in terms of multinational restructuring (Murray, 1985; see also Young and Mills, 1983: 43). In the early 1980s, it was estimated that some three-quarters of employees in the private sector of UK manufacturing worked for multiplant companies, a reflection of the rapid acceleration of takeovers and mergers in the later 1960s (Young and Mills, 1983: 50-1). 'In 1982, the 75 factories in London that still had more than 500 workers were all, with two exceptions, owned by multinationals. London's major offices, advertisers, accounting firms, financial institutions, insurance companies are again, predominantly owned by multinationals' (Murray, 1985). In London, the major locational shifts took place from the early 1960s; where there was considerable manufacturing growth in and around the central industrial conurbations of London in the 1950s, this trend had reversed in the 1960s.

The specific role of multinationals in closing subsidiaries in the inner London boroughs as part of a more global strategy began to be charted in the late 1970s (Brimson, 1979; Canning Town Community Development Project, 1977; Southwark Trades Council, 1976; Massey and Megan, 1980). Where manufacturing was still significant in some of these inner boroughs as late as 1966, the subsequent massive decline in dock employment and in related fields such as ship repair, chemicals, and electrical and mechanical engineering resulted from national restructuring in the face of international competition (Young and Mills, 1983: 75). The major impact came with the collapse of Dockland employment.

With the closure of St Katherine's Docks and the London and Surrey Docks between 1967 and 1970, the number of dockworkers fell by nearly one-half; by 1981 it had shrunk from 20,000 to 5,700; as the Royal Docks were shut down, and their functions transferred to Tilbury and elsewhere due to technological and transport changes (containerization) the dock function effectively disappeared (Falk, 1981; MSC, 1982; GLC, 1985). With it went a further 53,000 jobs in manufacturing; the space they once occupied has been taken over since 1980 by the London Dockland Development Corporation and the jobs created have been few compared to those which were lost. At the other side of the city, however, Heathrow airport and the growth of traffic has, to some extent (though hardly impacting on the same people), compensated for these changes.

These changes, now common knowledge, are charted elsewhere (GLC, 1985). The figure of 400,000 registered unemployed in March 1985 was expected to reach 545,000 by 1990 if present government policies continued — cuts in public spending, the application of new technology, and increased privatization. The result has been increasing polarization — of income, occupation, housing conditions, and life styles. The pattern of unemployment has been both structurally and spatially uneven, and it has been concentrated in the inner city. By 1982, the inner-city figure was 17 per cent (33 per cent in some boroughs) and 13 per cent in Greater London (Livingstone, 1982). Partly because of their concentration in such inner-city areas, it has disproportionately affected Black and Asian populations. In the mid-1980s, 25 per cent of Blacks but only 12 per cent of Whites were unemployed in inner London (Elliott, 1986: 105) though such figures too easily lead to stereotyping. Though general London unemployment figures have subsequently decreased, these basic disparities remain unchanged.

The industries that are growing are, like the financial sector, those connected to London's role as information city and centre of cultural production and ideological change. Inherently linked to that function is London's largest manufacturing sector in the late 1980s of printing and publishing, which employs 118,000 people (GLC, 1985: 172, 365). Sectors that are growing are those connected to this other aspect of London's role in the world-economy, though the number of jobs is also affected by new technology. (Printing is likewise New York's largest industry, see Sternleib and Hughes, 1988, as also Tokyo's.)

In the 1980s, as producers have moved to capture global markets, advertising has become more consciously transnational aiming at global markets and projecting global brand names. This has given an enhanced importance to advertising agencies, whether in the traditional centre of New York or increasingly, in that of London. In the three years up to 1987, by buying up competitors, Saatchi and Saatchi moved from being below the first hundred in Britain's 500 largest companies to being in the first ten and the largest agency in the world. (*The Times 1000, 1986-7*, which also indicates another advertising agency in the top-ten most profitable companies; in 1987 the two director brothers were each paid some £300,000 a year, and average staff remuneration across the industry was just over £20,000. The company had 150 offices round the world with 13,000 employees; *The Independent*, 2 October 1987.)

Increasingly, investment is put into changing consciousness rather than producing goods. According to the Institute of Practitioners in Advertising Annual Report for 1986, between 1976 and 1986, turnover in advertising agencies increased in real terms by 70 per cent compared to a mere 22 per cent in GNP. Employment in advertising in London (which accounts for almost 60 per cent of the total in the UK) numbers over 20,000 (GLC, 1985).

Advertising, printing, and a host of service industries (including employment agencies) grow hand in hand. The 1980s have seen a huge expansion of giveaway magazines, financed by advertisers and often handed out at underground stations. As commuters' time and space are limited, such magazines are limited by transport authorities so that each title is allocated a specified day and specified time at specified commuter stations; according to the Association of Free Newspapers' executive director, the number of such magazines has increased from 36 to 403 between 1976 and 1987 and of newspapers, from 169 to over 900 in the same period. Between them, they take over £400 million in advertising (*The Times*, 21 October 1987). Advertising City jobs, employment agencies, property, wordprocessing and secretarial services, health, or travel, the distribution is some 150,000 a week, paid for by the advertisers (*Nine to Five*, February, 1987). In addition, consumer magazines have grown in number and the managerial, administrative, marketing, property sales, and many other information functions of London have massively increased the amount of print and image-production media. The graphic and design aspects of this are discussed on p. 135.

But apart from advertising, cultural production in general has become increasingly important to London's international role (a valuable account is provided in GLC, 1985). This is not only in the conventional areas of the arts, literature, theatre, concert music, and the like but increasingly in the electronic media of radio, television, records, films, video with the cultural industries as a whole employing a quarter of a million people. In addition to printing and publishing on which this activity relies, the infrastructure also includes a substantial electrical engineering and telecommunications industry, the former employing 100,000. Culture, as the GLC's *Industrial Strategy* points out, is a major export from the UK, with over one-third of British books exported and a quarter of the world's records emanating from the UK. Some sectors of the printing industry such as newspapers have a much higher representation in London, with 40 per cent of national employment in newspaper and periodical production. This investment in cultural production is buttressed by other global institutions, the BBC, Reuters, Visnews, and others (GLC, 1985).

But as in other spheres of activity, the internationalization of control has been a feature of the 1980s (not least in newspapers with takeovers by Rupert Murdoch's News International, now owning one-third of British titles) and with cable and satellite transmission expanding, these developments are certain to increase.

Like banking and finance, the growth of the information and cultural production sector has had a similar effect on the generation of office space, extending its use particularly outside the central area to the south and the west. Here, the effects of privatization have been important. In 1986, the expansion of British Telecom led to the largest letting (of a quarter of a million sq ft in Croydon) of a completed building in Britain that year. As in the City, larger organizations have meant larger buildings with the average size of new building starts in 1985–86 increasing from 24,000 to 34,000 sq ft. Of all the new floor space completed in Greater London *outside* the central area since 1981, one-third was taken up by business services and within this, design, printing, and publishing companies occupied a significant amount throughout Greater London (other important sectors were industrials (15 per cent), computing (12 per cent), and banking and finance (10 per cent)) (Jones, Lang, Wootton, 1986).

ECONOMIC AND SOCIAL POLARIZATION

In 1977, Lapping suggested that 'the social polarisation of inner London is on the horizon'. Ten years later, a preliminary report by Townsend (1987) demonstrated the levels that this had reached. London had the largest concentration of unemployment of any city in the industrial world, with the mid-1985 figure of 400,000 probably underestimating the total by over 150,000. London's economy was seen to be 'in deeper crisis than it has been for a hundred years' (p. 12). The figures of the early 1980s had increased with unemployment averaging 25 per cent in the most deprived inner wards in May

1986, and being closer to 35 and 40 per cent in parts of Lambeth and Southwark (p. 83). In comparison, in the least-deprived wards of outer London, the average was just over 4 per cent.

As indicated in the previous section, minority groups and women were particularly affected. Unemployment and low incomes have also been reflected in a variety of disadvantages: overcrowding, single-parent families, shared household amenities, residential insecurity, and higher death rates. These indices of deprivation are equally manifest at spatial levels, with deprivation concentrated in central wards, the highest in the 1980s in wards in Tower Hamlets, Brent, Kensington and Chelsea, Hackney and Hammersmith, and Fulham. The least-deprived boroughs were those in outer London's Bromley, Havering, and Sutton. The mortality rate for the young and middle-aged in the most-deprived wards was double that in the least-deprived and large numbers of people were living below the Government's officially defined poverty line (Townsend, 1987).

The situation in London exemplifies, on a larger scale, the increased economic, social, and spatial polarization resulting from restructuring of the British economy as also the regressive tax and social policies that have characterized the 1980s. The disappearance of well-paid industrial employment has been accompanied by the growth of a proportion of more highly-paid service and business occupations at the top and the increase of low-paid, part-time temporary or casual work at the bottom. At the national level, income inequalities have increased because of both changed occupational and earnings patterns and taxation. According to Wicks (1987), between 1979 and 1986, net pay for the bottom 10 per cent of income earners rose by less than 3 per cent in real terms whilst among the top 10 per cent, it went up by a massive 21 per cent. Tax cuts have favoured the rich and in stark contrast, 'there has been a substantial increase in the number of people in Britain who are poor, close to poverty or hard-pressed' (p. 10).

Unemployment, a major force for inequality, has struck different sectors differently. Where typically, 1.5 per cent of professionals were unemployed (1985), the proportion was almost 13 per cent for the unskilled and 9.6 per cent for both partly skilled and skilled. Yet income and wealth statistics, as Wicks (1987) points out, hardly show the whole picture, affected as it is by access, or lack of it, to health care, education, shelter, or acceptance in the community. 'Statistics cannot document the contrast between a flat in a poorly built and badly maintained tower block in an inner city and a semi-detached house in the suburbs' (ibid).

In London, this polarization has been more stark, engendered by the accelerated processes of internationalization at the top and the bottom of the income and social scale. The internationalization of business and finance has meant that, increasingly, in particular sectors, 'world level' salaries theoretically and in practice determine rewards, irrespective of location. In the City, this has increasingly been the case as salary levels, traditionally much lower by international standards, have in 1980s begun to catch up with those in New York or elsewhere (Thrift, 1987b, from which the following is taken). Between 1985 and 1987, remuneration in the City of London (and by knock-on effect, in certain sectors outside it) has moved very significantly upwards. In 1985, the number of people earning £100,000 a year was relatively small (some 430 directors, top managers, and dealers, even though many of these made substantial sums on the market and in the lead up to Big Bang, selling out their firms and expertise to the new conglomerates). With international market skills in short supply, prior to deregulation, competition made high salaries (with additional bonus possibilities) the norm, a factor that the City was later to rue (*Economist* 7 July 1988). According to Thrift (1987b), salaries over £100,000 for a range of specialized jobs became 'fairly common' in 1987. For many in this category, these were also boosted by 'golden hellos', often up to £200,000, to tempt employees to new firms. Bonuses on a variety of other City jobs, from project finance manager (£36,000) to investment dealer (£66,000) took their respective salaries to £41,000 and £103,000 in May 1987. Directors' earnings were closer to £200,000.

Thrift (1987b) estimates that, with bonuses, about 4,000 people in the City were earning over £100,000 in 1986 (some, substantially more) and a further 10,000 to 15,000, between £50,000 and £100,000. In terms of total income (including rents, stocks, shares, etc.) these figures would be higher. In addition, the City had contributed considerably to the number of millionaires in Britain (20,000 in 1986), of whom at least 15 per cent were in the City.

At the other end of the scale, secretaries earned between £10,000 and £12,000 which, with 'perks' added another £2,000–£3,000. These, like the filing clerks, shorthand typists, and receptionists (at £5,000–£6,000 a year in 1987) were all women, as were the office cleaners at £3 per hour.

The internationalization of capital that is behind these City salaries, as also City rents and the boom in building, has been accompanied by the internationalization of migrant labour, which also helps to explain the stark contrasts of the City's immediate neighbour, the borough of Tower Hamlets. It is here where phrases such as 'the peripheralization of the core' and the comparison of 'Third-' with 'First-World' cities most applies.

According to *The Guardian* (15 December 1987) Tower Hamlets is 'one of the most deprived boroughs in the country' with half its residents earning less than £100 a week and with a large proportion unemployed. Traditionally the locale for refugees, Tower Hamlets, with its 10,000 Bengali residents (see p. 145) is the local result of labour migration working at the international level, the locale for sweated labour, high rates of unemployment, and homelessness.

Whilst the housing crisis is national, with the numbers of homeless increasing — at the close of 1987 — more rapidly in metropolitan areas outside London (Shelter, 1987), the major causes in the regions (household friction, marital disputes, and increasingly, mortgage arrears) are extensively exacerbated by the particular circumstances of London's role as 'world city'. Migration, both national and international, the impact of international investment in residential property, escalating producer-service salaries (and resulting gentrification), redevelopment, the international rental market, the disappearance of rental dwellings, combine with political policies of the deliberate contraction of state housing and other factors to result in extensive rises in house prices and, with unemployment and the growth of low-paid service jobs, a mounting problem of homelessness.

Whilst the full extent of this is difficult to measure, the number of households accepted as homeless by London boroughs more than doubled between 1975 and 1985 to over 27,000 (Conway, 1985; Shelter, 1985). The accommodation of homeless households in temporary 'bed and breakfast' hotels (generally consisting of one room per household) rose ten times from 890 in 1981 to 8,510 in December 1987, with the most recent figures showing a 36 per cent increase over the previous year. In particular London boroughs, the proportion of Black people in temporary accommodation is proportionately large (in 1985, in Haringey, 56 per cent; in Lambeth, 46 per cent). In Tower Hamlets, it was 90 per cent. Within walking distance, new one-bedroom flats for City executives in Docklands were selling, in late 1987, for over £100,000.

TOURISM AND THE WORLD-ECONOMY

As manufacturing has declined, tourism has boomed: the flow of goods going out has been replaced by people coming in. In 1982, the 20 million visitors to London were more than double the number two decades before; in the five years to 1987, they had increased by an estimated 11 per cent. Overseas tourists spent well over £3,000 billion in London in 1985 and another £733 billion on British carriers getting there. With tourist expenditure accounting for 8 per cent of London's GDP, it has become of increasing importance to London's economy. In all, it underpins a quarter of a million jobs, about half of them in hotels, restaurants, clubs, pubs, and catering (MSC, 1982; GLC, 1985; London Visitor and Convention Bureau, 1985; 1987).

Like other realms of London's economy, however, both the income and employment related to tourism depend on factors over which the city has no control: international exchange rates and the health, not only of the world-economy as a whole, but particular countries within it.

About 8 million of the 22 million visitors to London in the late 1980s came from overseas. Between 1983 and 1985, the largest number of these (35 per cent of the whole) has been from North America, 5 per cent more than from all the European Community countries combined. Particular sections of the London economy (most obviously the larger and smaller hotels, theatres, the culture industries, as well as particular shops) are largely dependent on tourism: 40 per cent of West End theatre tickets are bought by foreign tourists (many from the USA). After the bombing of Libya in 1986 and the consequent fear of reprisals on US citizens abroad, some leading hotel chains reported bookings down by 30 per cent.

As in New York or Los Angeles, however (Sassen-Koob, 1984; 1986), tourism, as one of the fastest growing service industries, is also the generator of many of the least-paid jobs. Whilst some tourist jobs are secure and well-paid, the large majority are part-time, casual, and filled by under- or non-unionized migrant labour. According to the GLC (1985), three-quarters of the workers are women, many of whom have come to Britain as migrant workers.

Hotel and catering is among the ten lowest paid industries in Britain and, as much of the work is done by non-unionized casual and foreign workers, their conditions have rarely been improved. Because of the difficulty in finding labour in the years of tourist expansion in the early

1970s, hotel employers recruited largely from abroad, from Italy, Spain, Portugal, the Philippines, and increasingly, from Latin America. A survey in 1971 showed that London's hotel industry had a higher concentration of immigrant workers than any other industry: 50 per cent of workers were born overseas (GLC, 1985). In a largely ununionized workforce (many of whom have little English) final decisions over work permits give additional powers of control.

As discussed on p. 106, the boom in tourism and the general global capital surplus has increased overseas investment in the hotel business. This has brought, according to the GLC, 'growing influence, both through direct ownership and management, of American union-busting techniques. . . . Holiday Inns took over a unionized hotel in Mayfair in 1984 and summarily ended its trade union agreement' (GLC, 1985: 470). This is yet another sphere where London space has become an arena for international conflict between capital and labour, though not necessarily between core and periphery.

Increasingly, hotel companies become multinationalized, plugging into the growing travel market (which constitutes 6 per cent of world trade) by offering a variety of travel, credit, or car-hire services.

Marketing the centre for tourism, therefore, has had a major impact on the built environment in general and housing in particular. What two generations ago were once private houses are now public hotels, many now, ironically, cashing in on the homeless situation in London; here, paid for by the local state, over 8,500 homeless households are accommodated in minimal 'bed and breakfast' accommodation, a figure that has risen by 2,000 over the previous year (*The Guardian*, 15 December 1987). In addition, in recent years, major public buildings have been turned into hotels: the headquarters of the Crown Agents in Millbank, St George's Hospital at Hyde Park Corner, and, at the time of writing, the likelihood is that the base of London local government, County Hall, will, following the Conservative government's abolition of the Greater London Council in 1985, follow the same fate (possibly combined with offices).

MAKING A MARKET IN CULTURE

In the economy of London, the promotion of tourism has increasingly become an alternative, from necessity rather than choice, to making and selling goods. London sells itself to the rest of the world — but it is also sold by others, coming in from the outside. This section considers London as a site for markets in education as another dimension of cultural production.

English, in its various forms, is now the lingua franca of the global economy. Three hundred million people speak it as either their first or only language and though there are well over three times that number speaking Chinese, English is the language of international business. Access to English, therefore, has become a major advantage, if not a prerequisite for operating in the world-economy. The acquisition of English has become increasingly commodified and in a stroke of serendipity, the market has grown in England and especially, in London. In the mid-1980s, teaching English to foreigners was said to be a £250 million business.

In the last twenty-five years, private language schools (largely teaching English but other languages as well), in bringing students from all over the world, have had a significant impact on the internationalization of Central London. This has helped change its social and cultural composition and, through legal and illegal employment, has also subsidized its low wage economy. They have also had an influence (through accommodation) on the domestic economy of local residents. In 1951, there were eleven private language schools in Central London offering English language; by 1962, this had already increased fivefold (King, 1962). Between 1970 and 1985 the number, both in Central London (over one hundred) and in London as a whole, doubled again, with particular concentrations of schools in the central area, the north-west (Hampstead, Camden), the north (including Highgate, Islington), and south-west (including Kensington) (Table 5.6).

The language schools are only one source of bringing international students to London. With well over a million students in the world pursuing courses outside their own country, the market in higher education has become big business. Yet historical, cultural, geographical, as well as economic factors, still help determine which students from what countries go where.

In 1983–4, of some 50,000 overseas students in Britain, over half were from colonial (Hong Kong) and ex-colonial countries, 40 per cent from six countries alone (Hong Kong, Malaysia, Nigeria, Singapore, Brunei, and Kenya). Of the other half, the growth in numbers was largely from the Middle East and the Gulf, the Far East, the European Community, Africa, and more recently, the USA. Proportionately, over a half were from Asia, about a quarter from Africa,

Table 5.5 Private language schools in London, 1970–85

	1970	1985
Central area	44	112
North-west	4	32
North	13	29
East	—	4
South-west	15	16
South-east	9	13
South	6	17
	91	223

Source: Post Office/British Telecom, *London Telephone Directory* Yellow Pages, 1970; 1985.
Note: There is some repetition of entries between central London and the north-west area.

15 per cent from Europe, and 10 per cent from the USA. A much smaller proportion were from South America, and these, from Brazil and Mexico. About 30 per cent of all were in London (British Council, 1984; 1988). (These figures all relate to public institutions.)

Whilst the total of international students in both private- and public-sector institutions is hard to assess, between 1982 and 1986, numbers in the public sector (universities, polytechnics, and colleges of higher education) in London have numbered about 16,000 to 18,000. Since the Thatcher government's back-door privatization of previously public institutions, including higher education, international students have increasingly been seen as a source of income, a critical contribution to keeping some courses afloat.

Income from the fees of international students in 1986 formed 5.3 per cent of the total income in British universities, though in individual cases, it reached almost 15 per cent. In the University of London it was some 6 per cent and in City University, almost 10 per cent (*Times Higher Education Supplement*, 2 October 1987, p. 4). In certain institutions, particularly oriented to international studies and with large graduate programmes the proportion of international students is higher than elsewhere.

More recently, attention has shifted from the economic impact of foreign students on institutions to their impact on regional economies, research significantly undertaken in one of Britain's ex-colonial port cities, most badly affected by imperial decline: in the mid-1980s, Glasgow hosted 25 per cent of all overseas students in Scotland. The authors of this study concluded that the impact of overseas students on the Scottish economy was 'very substantial' reaching far beyond

127

their host institutions, and that there may be similarities in this pattern throughout the UK (Love and McNicoll, 1988).

In the last few years, the total UK figure has increased and, for a variety of reasons, the country composition has changed. The fall of the price of oil on the world market has affected oil-producing countries (e.g. Nigeria, Venezuela, and the Middle East), and the world-debt crises in 1982 and 1983 in both African and Latin American countries, by restricting foreign exchange available to overseas students, has cut down numbers. Political crises (in Iran, Lebanon, or between Britain and Libya) likewise affect numbers. Not least, economic crisis in higher education in the UK since 1983 has also resulted in more intensive recruiting policies to boost institutional income from foreign students' fees.

Hence, by 1985–6, total numbers had increased to 63,500, though with Commonwealth senders contributing under half. The richer, newly industrializing countries such as Hong Kong and Malaysia, with over 13,000 students in all, are the leading customers for British higher education but between 1983–4 and 1985–6, American students in public higher education institutions in Britain have increased by 60 per cent (to 4,600). Though numbers are still small, the greatest percentage increase (65 per cent) in the last two years has been in students from China (675 in 1985–6), Italy (39 per cent), France (30 per cent), Canada (28 per cent), and Australia (22 per cent) (British Council, 1988).

Whilst many factors affect the locational decision of students wishing to take their degrees outside their own countries, in many of the newly industrializing countries of the world (e.g. in East and South-East Asia), higher education institutions cannot accommodate the numbers wanting higher education.

In this world market in higher education, the UK and the USA are, to varying degrees and for different reasons, in direct competition. Where decisions are made on the grounds of cost and exchange rates, US institutions have significant advantages: figures show that of almost 330,000 foreign students in the USA in 1986 (including 25,600 from Taiwan, 21,600 (Malaysia), 20,000 (China), 20,000 (Korea), 18,000 (India), and 15,000 each from Canada and Japan), over 23 per cent had funding from their US college or university, 26 per cent from the US government, 21 per cent from their own employment, 8.6 per cent from their governments or university, and only 27 per cent from 'foreign private' sources (Institute of International

Education, 1986a). In the context of these figures, UK reliance on linguistic, cultural, or historical connections and significantly less on financial aid raises questions about the realism of current recruitment policies, irrespective of the cultural politics involved.

Where the interest of private schools is in language, the interest of American colleges is in culture. Since 1970, London (and Britain) have become an educational off-shore island for the USA, extending American branch-plant education abroad.

As American college education has become more consciously international, students have been encouraged to spend more time away from home. In addition, private educational institutions have seen the 'study year abroad' as a factor affecting their competitive position. Hence, universities and colleges in the US (often in collaboration with those in the UK) have sponsored study programmes abroad, many in the UK and many based in London.

Starting slowly in the 1950s, the number of these programmes accelerated rapidly from the 1970s (Figure 5.2). Whereas there were only some 30 in London in 1970, out of a total of about 650 listed in the Institute of International Education's *Study in the UK and Ireland* (1986b), at least 230 were in London. Western Europe, with almost 75 per cent, had the highest proportion of American students studying abroad: within Europe, the UK, with 13,700 in 1986, had the highest proportion of the total (28 per cent), twice as many as were in France. These students make a substantial contribution to the international and specifically American presence in London (see p. 126).

Private American educational establishments, more familiar with the operations of the market than their British public-sector counterparts, have steadily built up business in the world education market in London. These exploit its cultural space, international location, public resources, and the growing pool of students coming in from all over the world. Their orientation, as their literature, is global. One college 'is defined by its international constituencies, its location at the global crossroads . . . London is probably the world's most international city . . . housing nearly one-fifth of (Britain's) population. Every student — British, North American, Asian, Middle-Eastern, Latin American — works at the core of global and national affairs'. Other colleges, 'with Buckingham Palace and The Royal Mews as close neighbours', make particular use of the business connections, specializing in graduate programmes on marketing, management, and computer-resource management. 'London is an ideal place to study

129

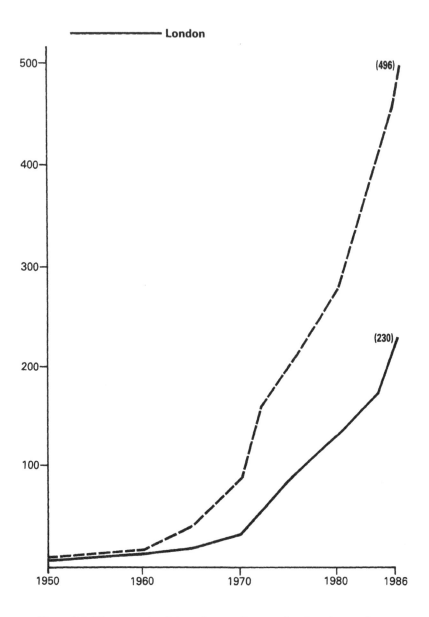

Figure 5.2 The growth of American college and university study
programmes in the UK and London, 1950–86

business and international relations. It is the center of the Eurodollar and Eurobond markets, as well as the home of Lloyds of London. Almost every major bank in the world is represented here'.[5]

Many American universities have set up permanent bases in London.[6] In 1985, under threat from the Thatcher government's withdrawal of funds from higher education, the University of London's Bedford College was forced to abandon its prestigious central location in Regents Park. One building was sold to a Middle-Eastern buyer; others were taken over by Regent's College, a base for private American colleges in the UK.

These legitimate educational and business institutions have, since the Government's liberalization programme and the Department of Education and Science's decision to discontinue inspecting private educational establishments, been followed by the growth of dubious (i.e. non-recognized) organizations offering (for considerable money) bogus degrees. It is a comment on the perceived status of London in world higher education that 57 out of 150 of such organizations in Britain incorporate 'London' in their designation (J. Belcher, QMC, University of London, letter to *The Guardian*, 3 December 1987; Council of Europe, 1986).

As fees have risen and scholarships and grants for students become scarce, studying in London for students from middle- or low-income countries tends to become increasingly a privilege of the wealthy (if privately funded) or for the relatively poor (if supported by governments or scholarships). (In 1986, rent, fees, and sustenance for one year's graduate study for an overseas student in architecture or planning cost between £9,000 and £10,000.)

PRODUCER SERVICES IN DESIGN: LINKING CORE AND PERIPHERY ENVIRONMENTS

In the operations of the capitalist world-economy, one function of institutions at the core is the incorporation of peripheral countries through the remodelling of environments, integrating countries into an increasingly global system of production and consumption. Initially, the functions are concerned with extraction and production: the exploitation of natural resources, oil exploration, refineries, and the setting up of the basic plant to do this. Simultaneously, a transportational infrastructure is required to channel the products into the world-economy: harbours, docks, highway systems, airports. A further stage is urban development in terms of transport systems, utilities, housing for labour, a managerial class, urban management. At a still further stage, is the finishing of the built environment in the form of internal and external design, materials, and appearances.

The responsibility for these tasks lies principally with the 'environmental professions' located in major world cities: consulting engineers, civil, mechanical, geotechnical; urban development planners, architects, urban design consultants, interior, product, and graphic design specialists – the image-creating professions. At the periphery, the functional requirements of production (surveying, extraction, processing) overlap with the functional requirements of consumption (urban planning and the construction of housing and retail outlets).

These services have long been established in world cities but in the last two decades, their production and export have become increasingly important to urban and national economies at the core. In the new international division of labour, the export of services takes over from the export of goods (Cohen, 1981; Sassen-Koob, 1984; Thrift, 1986a). In addition, developments in the world city itself have, as indicated above, raised the demand for these services. According to Gutman (1988) and Knox (1987) in the USA, architecture is the fastest growing profession with the number of architects trebling between 1960 and 1980. Much of this growth is tied to the corporate economy but it is also from overseas activity with the export of design services accounting for some 15 per cent of contracts in 1983. These developments can be illustrated with data from the UK.

Of some 600 firms listed in the *Consulting Engineers Who's Who and Yearbook*, 1982–3, the offices of about one-third (including most, though not all of the largest firms) are located in London (about 110) or the immediate London region (about 105): i.e. Reading, Croydon, Epsom, Ware, etc.). Of those with London addresses, 60 per cent are evenly divided between W1 and SW1.

However, the initial impression that London-region consulting engineers undertake operations on a worldwide basis needs qualification. Though much more detailed data would be needed on the size of contracts, a rough count of the frequency with which countries are mentioned where contracts have been carried out suggests that not only are many of these in the middle- and low-income countries of the world-economy but that earlier historical and political connections are more important than geography or economic factors alone.

132

A similar impression arises from details on architectural firms. The Royal Institute of British Architects (RIBA) lists 1,450 firms in its *Directory of International Practice* (1980), with experience of overseas work in the period 1971–1981. All but some 5 per cent were based in the UK. Of these, some 20 per cent (about 280) including most, though again not all of the major practices, were based in London. Of these, 55 per cent were in or close to the West End (70 in W1, 50 in WC1 or WC2). The figures suggest (see Table 5.6) that in the 1970s for both groups, most contact was with the Middle East (especially Saudi Arabia and, in the past, for architects, in Iran); Africa was also important (especially Nigeria and Libya). For architects, Europe was also important, especially France (114) though other European Community countries were mentioned 30–50 times. South Asia (India and Pakistan) and the Far East (Hong Kong and Singapore) were relatively less important. By comparison, far fewer British consultants and firms had contacts with South America (and these, mainly with Brazil), Central America, and the Caribbean (most frequently mentioned architectural contacts were with the Bahamas (18) and Jamaica (17)). For architects, South-East Asia (81, especially Malaysia and Indonesia) were important though there is far less contact with other countries there (e.g. Japan, 10 and Korea, 5).

Table 5.6 Overseas contacts of UK engineering/architecture consultants: by continent/region, 1971–81

	Engineering	Architecture	Total
Middle East	130	630	760
Africa	147	442	589
Europe	32	546	578
South Asia	23	85	108
Far East	35	66	101
South America	25	69	94
Central America/ Caribbean	16	35	51

Source: RIBA, 1981; *Consulting Engineers Who's Who and Yearbook*, 1982–3.
Notes: *Figures are approximate and refer only to countries mentioned.

More recent data for architectural firms having experience of overseas work (including RIBA members in firms based overseas (RIBA, 1986)), based on a larger number of member firms (2162, of which 18 per cent were overseas and of the 1769 UK firms, 11 per cent of which were in London) indicates the following number of contacts between 1975–85 (see Table 5.7).

Table 5.7 Overseas contacts of UK architectural consultants, by continent/region, 1975–85

Europe	895
Africa	876
Middle East	736
South-East Asia	280
North America*	257
Far East	201
Central America/Caribbean	198
South Asia	161
Australasia*	145
South America	126

Source: RIBA 1986.
Note: *Not included in Table 5.6.

In Europe, the most frequently cited countries are France, Eire, and Spain; in Africa: Nigeria, South Africa, and Egypt; in the Middle East: Saudi Arabia, the UAE, and Qatar; both the USA and Canada are cited; in the Far East: Hong Kong and China; and in the Caribbean/Central America:Trinidad and Tobago and Jamaica. Again, South America figures far less prominently than other regions and here, Paraguay (25) and Venezuela (19) are the most frequently cited countries.

Three generalizations can be made: the first concerns the very wide economic, ideological, and cultural influences exercised by specific professional services; the second, that such influences are still predominantly from core to periphery, though the move of Europe to first place and the increasing importance of North America are worth noting; the third suggests that the areas of influence are still determined by important historical and political criteria. What needs examination is the hypothesis that the capitalist world-economy is carved up into different operational spheres and managed by specific countries at the core: equally important to examine are the changes that these figures suggest have taken place in recent years and their effects.[7]

For example, the emergence in recent years of Turkish, Yugoslavian, and South Korean contractors, using low-cost labour, has made inroads into Middle-Eastern markets. These, with declining oil prices, the Gulf War, and the completion of many earlier projects, have declined in importance. Yet the growth in contractors from other countries, particularly Turkey (as a Muslim country), has been significant, as has the increasing importance of indigenous contractors. Likewise, in 'traditional' English markets like Nigeria and East Africa, French companies have been making inroads while in South-East Asia, although future prospects seemed good, fierce competition comes from Japanese and South Korean contractors. And in what were once 'developing countries', contractors enter the industry backed by government finance and, as with Korea, offering control of their own labour (Seymour, 1986). These are some of the factors explaining the shift in the geographical focus of UK firms in recent years.

The third of these environmental professions provides design services. An increasing output of multidisciplinary design consultants, covering product, interior, and graphic design, stems from international clients: annual reports for foreign banks, publicity for project funding in 'developing countries', promotional material for Middle-Eastern companies, or space planning for multinationals locating abroad. Unlike engineering, planning, or architecture, however, the full professionalization of design consultants (in terms of a single professional institution, etc.) has not yet occurred; hence, comparable data on overseas operations are not available.[8] In the last two decades, however, design has been a major growth industry in the capital.

The number of graphic- and advertising-design firms more than doubled (to 915) between 1970 and 1985 while 'design consultants' (unlisted in the Yellow Pages in 1960) increased by over 70 per cent (Table 5.8). Interior designers (again, not listed in 1960, or classed with 'interior decorators' and less conscious of professional status) increased by 84 per cent. The growth of these activities is closely allied to the developments outlined in earlier sections: the development of London as an international information city with printing as the major industry (as is also the case in New York and Tokyo): everything that is printed is designed; the growth in advertising and marketing; the increase in building and redevelopment and, not least, the expansion of the international market in residential property.

In recent years, the glossy brochures promoting house sales in potentially upmarket areas have increasingly used artists' water-colour illustrations rather than photographs; by removing cars, cracks in the brickwork, or rundown surroundings, these project property in a more

Table 5.8 'Environmental' professions: architectural, design, and related firms in London, 1970–1985

(London)	Central		N. West		North		East		S. West		S. East		South		London Total		
	1970	85	70	85	70	85	70	85	70	85	70	85	70	85	1970	1985	(1960)
1. Architects	618	547	127	245	245	300	80	76	290	283	100	92	93	152	1553	1695	(1180†)*
2. Town Planning	8	18	3	8	3	8	—	2	—	3	1	1	2	4	17	44	
3. Artists (commercial, industrial)	180	113	25	17	30	17	24	15	34	17	23	22	14	18	330	219	(250)
4. Art and layout services	*	38	*	4	*	3	1	3	6	10	2	11	3	14	12	83	
5. Design consultants	262	337	14	125	67	112	7	26	60	120	40	34	19	58	469	812	*
6. Designers (advertising, graphic)	300	372	—	119		112	7	38	21	120	40	49	9	105	377	915	*
7. Designers (exhibition)	4	5	—	5	1	3	2	1	3	15	3	1	1	3	14	33	
8. Designers (TV, film, theatre)	3	5	4	2	1	1	—	1	—	4	1	1	1	1	10	15	
9. Interior designers	186	193	*	126	154	93	—	28	—	98	—	25	—	63	340	626	
A Sub-total	1561	1628	173	651	501	649	121	190	414	670	210	236	142	418	3122	4442	
10. Art galleries and dealers (private)	235	248	7	61	28	37	5	22	49	42	8	12	11	26	343	448	(126)
11. Antique dealers	830	729	71	340	200	248	43	81	550	312	140	147	108	171	1942	2028	(706)
12. Antique repairers, restorers, valuers	50	62	9	38	31	42	6	17	51	38	10	19	8	35	165	251	
B Total	2676	2667	260	1090	760	976	175	310	1064	1062	368	414	269	650	5572	7169	

Source: Post Office (1970) and British Telecom (1985), London Directories
Note: †Includes architects and surveyors
*See note 9

flattering (if less honest) light. For international clients wanting houses ready to occupy, or let, interior designers need furnishings and paintings for the walls; status-giving 'antiques' are bought to give places the needed cachet. As Britain is an old country, and London the centre of the trade, a large number of firms have arisen engaged in recommodifying obsolete objects, reconstituting their histories, investing them with meaning, and legitimizing their authenticity. These are some of the reasons for the more than threefold increase in private art galleries and dealers between 1960 and 1985, and likewise, the threefold increase in antique dealers over the same period (though this was largely in the 1960s, from 706 to 1942 by 1970; the number, just over 2000 in 1985, having only slightly risen from 1970).

Oil-surplus funds are invested in mansions, whether in the UK or in the Middle East, decorated with specially commissioned interiors. The complete internationalization of the art market and the function of art objects as receptacles for storing capital has also given increased and critical importance to the international auctions and ownership of the principal houses. Sotheby's was acquired by American owners in 1983, since when its turnover has increased fourfold. New, internationally generated projects provide work for artists and photographers of all kinds. (The controversy over the proposed Canary Wharf skyscrapers led one architectural firm involved to hire photographers to site themselves in prominent locations around London one morning to take simultaneous photographs of a helicopter, hovering at the height of the projected highest tower. The photographs were then used to produce a simulated 'environmental impact assessment' of the project.)

Taking all these 'environmental' activities together (i.e. architects, design consultants, interior design, art galleries, etc.), there has been a 40 per cent increase in the number of firms engaged in them between 1970–85, though the increase has affected different areas differently.

It has also helped to change their social composition and character. Prevented by higher rental uses from expansion in the central area (where numbers have remained virtually the same), firms have expanded into districts both adjoining and distant from the central area. Design consultants, though still expanding in W1 and WC1 or WC2, have also expanded into the areas north, south-west and south of Central London. In 1985, there were six times as many graphic and advertising design firms listed in the south-west (i.e. Richmond, Kingston, Twickenham) as in 1970 and ten times as many south of the river as fifteen years before. Along with the changing social composition of areas, these changes have added to the 'gentrification' of areas and are used by real-estate agents to modify the image of previously downmarket areas.

As Table 5.9 shows, the significance of the north-west (including Campden, Hampstead) in these activities has enormously increased, with a fourfold growth in firms in fifteen years. Also worth noting is the 140 per cent increase in the south, still (with the exception of particular areas) a relatively 'downmarket' area in the 1960s.[9]

Table 5.9 Changing locations of 'environmental' professions and firms in London, 1970–1985

1970			*1985*
Central	2667	2676	Central
South-west	1064	1093	North-west
North	760	1062	South-west
South-east	368	976	North
South	269	650	South
North-west	260	414	South-east
East	175	310	East

Source Post Office (1970) and British Telecom (1985), *London Directories* (Yellow Pages).

These developments are used by a coalition of firms to promote their interests and, not least, to increase property values. The February 1987 edition of the monthly glossy magazine, *Southside* (delivered free to houses in the region) sees the 'psychological barrier of the river' as having disappeared, translating the South Bank as 'Rive Gauche'. It draws attention to the fact that 'internationally renowned designers and decorators whose clients include the King of Saudi Arabia and the Prince of Wales now live south of the river' (e.g. David Hicks International, Liberty of London Prints, Charles Hammond, etc.). Art, one of the oldest symbols of status, valorizes location through association. New art galleries (fifteen since 1970, but especially growing since the early 1980s) are mainly in 'upwardly mobile residential areas' (e.g. Wandsworth, Battersea, Clapham, and Southwark). Articles feature books, arts, new businesses, finance ('How to read an annual report' and 'The tide of businesses, flooding south'). One column discusses ' "In" roads' (areas going upmarket). Advertisements are for the many new restaurants and the newly commodified activities such as 'Children's party entertainer' or for services that clear up after dinners and parties. Much of the financial support comes from the many new estate agents who advertise in its pages.

As a major mechanism in the process of ideological and cultural change, projecting images, creating markets for goods, and moulding the non-western world in the cultural image of the core, the international operations of the environmental and especially design professions deserve more research attention than they have so far received.

SOCIAL, ETHNIC, AND SPATIAL RESTRUCTURING

The question of social, ethnic, and spatial polarization in London is too important an issue to be dealt with either briefly or on the basis of inadequate data. Since the original version of this chapter was completed in 1983, significant changes, as well as socially charged events have occurred, which make the construction of an overall picture, as informed by propositions from world-city literature, both imperative and urgent. What follows relates firstly to aspects of the international division of labour and their social and ethnic effects on London.

A historical trend towards a theoretical 'global labour market' in which labour moved freely round the world ought to be a testable proposition: in reality, however, within a worldwide division of labour, the network of cross-national movements is much more complex (Petras, 1981). Geographic proximity, cultural affinity, historic, political, and economic networks, and the role of the State result in world-city populations that have very specific ethnic and national characteristics: London is no exception.

Within an overall situation of population decline in Greater London, decline in the inner city (a continuing process since 1911) has been far more rapid than in the city as a whole, reducing by 5 per cent between 1951–61, 13 per cent (1961–71), and 18 per cent (1971–81) (Hamnett and Randolph, 1982). The ethnic and social characteristics of inner and outer London have also undergone major changes in the last two decades.

If it is obvious to mention that state policies regulate and define the conditions of boundary crossing by migrant labour (Petras, 1981) it is done so to emphasize that recent UK census-statistics data have been published in categories, not to test hypotheses about a theoretical, global labour market but rather, to monitor immigration on the basis of race. Although only one census question relates to birthplace, ethnic origin, or nationality, namely, that on the country of birth, the resultant data on countries of birth are classified broadly according to five categories: whether residents are born in the UK, outside the UK (i.e. Eire), 'Old Commonwealth' (Australia, Canada, and New Zealand), 'New Commonwealth' (i.e. countries in Africa, Asia, the Caribbean, and the Mediterranean) and 'Foreign'. As Lee (1978: 10) comments, 'The term (New Commonwealth) is widely regarded as a euphemism for the coloured population, although the inclusion of groups such as Cypriots within this term demonstrates the inaccuracy of this interpretation'.

Of the total Greater London population in 1981, about one-fifth were born outside the UK; of these, the largest single group were from the Irish Republic (some 200,000) apparently confirming Petras' (1981) point that, among major labour movements today, the largest are between contiguous states (cf. Mexico and Canada in relation to the USA).

In contradiction to this, however, is the combined total of those born in 'Old Commonwealth' (OC) (37,000) and 'New Commonwealth and Pakistan' (NCWP), a figure of some 632, 000. In short, over half of all migrants in London born outside the UK are from 'OC' and 'NCWP', almost twice as many as the 371,000 'Foreign' residents. Irrespective of political/national divisions the numbers are indicated in Table 5.10. The data clearly confirm, if confirmation is needed, the degree to which the composition of labour is determined by historically and politically specific influences, a point further confirmed by trends between 1971–81. In a situation of overall decline, the only birthplace group (accepting Census categories) to have increased were those from NCWP, whose numbers rose by 35 per cent (105,000) so that households where the head is of so-called NCWP origin made up 15 per cent of Greater London's population in 1981 (GLC, 1983b). A large part of this increase was accounted for, not by wage or economic factors but by political events in East Africa resulting in a fourfold increase in persons born in Uganda, a threefold increase in Tanzania, and a doubling of those born in Kenya.

Though figures of those born in Europe partly lend credence to the notion of a 'free labour market', historically specific explanations are also necessary. Thus, of groups over 10,000, those born in Italy (31,000), Spain (21,000), and Portugal (11,000) suggest the migration of labour from low- to higher-wage zones and, as discussed on p. 139, this has in fact been the case (see also GLC, 1985; 469); similar

140

Table 5.10 Greater London. Usually resident population: country of birth by continent/region, 1971 and 1981

		1971		*1981*
Total population		7,500,000		6,600,000
Born outside the UK		1,069,000		1,200,000
Europe		401,000		382,000
incl. Eire	198,000		199,000	
Asia		229,000		296,000
Africa		92,000		170,000
Caribbean		169,000		168,000
Mediterranean		64,000		69,000
(Gibraltar, Gozo,				
Cyprus, Malta)				
North America		36,000		32,000
United States	24,000		22,000	
Canada	12,000		10,000	
Middle East				30,000
incl. Iran	4,000		12,000	
Australasia		28,000		27,000
Australia	20,000		16,000	
New Zealand	8,000		11,000	
South & Central America		7,000		11,000
USSR (not incl. above)		11,000		7,000

Source: OPCS, 1973 (Table 14); 1982 (Table 51).
Note: *Not available.

figures for West Germany (30,000), Poland (26,000), or outside Europe, South Africa (16,000) require other explanations.

Whereas overall, the percentage of European residents has decreased between 1971–81 by 8 per cent and those from 'the rest of the world' by 7 per cent, residents born in Middle-Eastern countries other than Iran, insufficiently numerically significant to be listed separately in 1971, numbered 18,000 in 1981.

Two facts emerge: the closer one moves to the centre of Greater London, the smaller the proportion of the population born in the UK. Second, despite, or because of an almost 10 per cent decline in population, all of Greater London's thirty-two boroughs became more cosmopolitan between 1971 and 1981 (i.e. the percentage of residents born in the UK decreased) (GLC, 1983a). Central boroughs, with only 60 per cent of their residents born in the UK, and 7 per cent in Eire, have 11 per cent NCWP residents, and 21 per cent from elsewhere abroad. Kensington and Chelsea, which experienced the highest loss of population of any English borough between 1971 and 1981 (26 per cent) has also the highest proportion (38 per cent) of non-UK-born residents, 10 per cent of whom are from NCWP origins. Similarly, the City of Westminster, with 34 per cent of non-UK-born residents (12 per cent NCWP) and Camden (which includes Hampstead) have 22 per cent 'foreign' residents (16 per cent NCWP). These trends certainly increased in the 1980s.

In 1987, 172 different languages were spoken by children in the schools of the Inner London Education Authority, an increase of 25 since 1985. Almost 23 per cent of pupils do not speak English at home. The number of Bengali speakers (the most common language apart from English) had tripled to 16,976 between 1981 and 1987 and Vietnamese speakers had risen sixfold in the same period. Other common languages included:

Turkish	4,495
Chinese	4,325
Gujerati	3,930
Urdu	3,808
Spanish	3,229
Punjabi	3,200
Arabic	3,067
Greek	2,596

The most commonly spoken language varied with the area, from Bengali in Tower Hamlets (East London) to Spanish in Chelsea (West London). In Holland Park Comprehensive School, West London, 50 languages were spoken (*The Guardian*, 10 November 1987). These and other data on the present social and ethnic composition of London's population are best understood within the framework suggested in Chapter 4 (Table 4.3). This implies an examination of the data over a more long-term (say, 50-year) period: whilst the following comments are speculative, they suggest issues for further research.

In the first half of the twentieth century, London's role in the old international divison of labour was still largely that of imperial city. It was the control centre of a colonial mode of production and, as such, its largely indigenous population was characterized by a social and class composition principally influenced by this role. This social composition also included a significant working-class element·based on a substantial amount of manufacturing: it was also both spatially as well as physically expressed in the built environment.

With the disintegration of the colonial mode of production and old international division of labour from the 1950s, the transition from the role of imperial city to world city was well underway. As in the emergent phase of other world cities (New York and Los Angeles), movement in the new international division of labour was from poor countries in the periphery to rich countries at the core. In the case of the UK, however, historical and political links with the old colonies (the Commonwealth) ensured that, along with geographically-determined migration from adjacent Eire (Petras, 1981), international migration in the 1950s was from the low-income countries of the Commonwealth — the West Indies, India, Pakistan, and subsequently, Bangladesh (as Hudson and Williams (1986: 134) point out, the difference with European labour migrations was that in Britain, migration took the form of permanent settlement as citizens rather than a temporary movement as 'guest workers').

Initially, this labour was largely concentrated in inner-city areas, especially in London, as 'replacement labour' engaged in poorly paid service occupations, transport, and 'dirty' industrial jobs (Hudson and Williams, 1986: 134) or in propping up industries that were becoming economically obsolescent in global terms.

Since the 1960s, the increasing specialization of London as world city, with its specific functions in the world-economy (and the characteristics that have accompanied these) has had various important effects on the social, ethnic, and spatial composition of its population. At the lower end of the social hierarchy, the collapse of industry (as well as a variety of other factors) has left the country- and culture-specific Black and Asian migrant labour, drawn from the poorest countries in the world, largely trapped in the run-down parts of the inner city. Whilst some 16 per cent of London's total population is Black or Asian (Elliott, 1986: 16), this proportion is much higher in specific inner or near-inner London boroughs (e.g. Brent, 33 per cent, Hackney, 28 per cent, Newham, 27 per cent, and Lambeth, 25 per cent, compared to outer London boroughs such as Hillingdon, 7 per cent, Richmond, 6 per cent, Kingston, 6 per cent, or Croydon, 13 per cent).

At the top end of the social hierarchy, however, London's enhanced role in the international division of labour as world banking, financial trading, as well as 'global control' centre, has had an equally important impact on its social composition.

The first effect is the concentration of a highly paid elite, principally (though not entirely) indigenous. The second effect results from the banking, financial, and trading function. As London is the principal financial centre in the capitalist world-economy, it requires the presence of representatives of the world's richest nations, most specifically, Japan and the United States. In addition, its role (like that of other world cities) as centre for international investment and capital accumulation brings in representatives of the world's richest states (Saudi Arabia and other gulf states, as well as the USA and Japan) as also, rich members of poorer countries. Both the rich at the top and the poor at the bottom have significant effects on the institutions and culture of the city, as also on the extent and expression of economic and social polarization and the potential for social, racial, ethnic, and political conflict (and also, co-operation).

Estimates of the number of Japanese in London range from 8,000 families, perhaps 30,000 in all (*The Times*, 20 October 1987) to 50,000 (*The Financial Times*, 25 July 1987). Entries under 'Japan' and 'Japanese' in the London Telephone Directory have increased sixfold (to 60) since 1960. These include a wide variety of shops, commercial and social organizations, translation services, cultural institutes, banks, services, newspapers, and schools related to the Japanese presence in London.

Similarly, in the context of the increase in banking, securities trading, multinational control, tourism, education, military presence, capital accumulation, cultural diplomacy, commercial activity and other of the functions mentioned above, it is likely that the number of Americans has increased beyond the 22,000 listed in the 1981 census (according to *Statistical Abstracts of the United States*, Washington, 1986, there were 186,000 American residents in the UK in 1986, including servicemen and dependents). The extent of both commercial and private business activity accounts for over 90 American law firms in London, largely (40) in W1 and SW1 and also (25) in EC1, EC2, EC3, and EC4. In addition, however, there are three American schools, colleges, a church, property-finding services, a women's club, a chamber of commerce, a newspaper, an estate office, specialized financial, insurance, tax, and legal services, banks, as well as the main consular, diplomatic, and commercial organizations, and a broadcasting company.

The effect of the world city, therefore, is to bring representatives of the richest and poorest countries in the world into the ambit of each other. It is ambit, rather than contact, because the historical economic, class, spatial, and built-environment divisions, between the West End and East End, which divided rich and poor in the Imperial City, continue to keep both groups apart.[10]

Thus, in 1981, both Bangladesh and the USA each had 22,000 of their citizens living in London, of whom 17,000 and 14,000, respectively, lived in the central area. Their spatial distribution, however, was to largely keep them apart

	Kensington and Chelsea	City of Westminster	Tower Hamlets
Bangladesh	256	953	9,808
USA	3,332	3,340	124

Two hypotheses might be subjected to more careful research: first, that within the wage zones of a global hierarchy of labour reward, migration from low-wage zones at the periphery, monitored by specific historical and political circumstances, results in movement to the lowest wage zones at the core; second, this movement is reflected in the social and ethnic structure of residential areas in world cities. Put another way, to what extent are global inequalities reproduced in the spatial distribution and spatial inequalities of populations in the world city?

These, and the data discussed earlier, suggest a situation where the centre of the world city becomes, like 1930s Shanghai, an international enclave whose space, social relations, and politics increasingly depend on decisions made outside national boundaries.

CONCLUSIONS

In the four years since the first version of this case study was completed, the trends and tendencies identified (King, 1984b) have very markedly increased. They are tendencies not simply towards an intensification in the 'internationalization' of London but its internationalization under the particular conditions, and with the particular outcomes, determined by the interests of international capital and the particular countries where these interests are principally based.

The strongest impression that emerges from the study (and this is confirmed by Sassen-Koob's work on New York; personal communication, July, 1987), is that the world city is increasingly 'unhooked' from the state where it exists, its fortunes decided by forces over which it has little control.

Increasingly, the city becomes an arena for capital, the site for the specialized operations of a global market. Forced to compete with its major international rivals, obstacles to that competition are, dependent on state policies, progressively removed. It is here where the interests of local populations are directly in conflict with, and are sacrificed for, the interests of international capital; and it is in this context that the abolition of the Labour-led Greater London Council in 1986 is to be understood.

Another analogy is that of the colonial city, or rather, the old Treaty Ports of China. Laid bare in the international market place, the centre of the city is colonized by capital. The processes are embodied in language and represented in space; 'concessions' are granted to foreign powers; 'enclaves' are created for tax-free economic activities; and to protect themselves from the 'natives', the representatives of national and international capital retreat into 'compounds' bounded by high fences, locked gates, and patrolled by state police or the security guards of private armies. In return for its percentage, the State maintains law and order, invests in the police, and provides the coolies and the social and physical infrastructure.

These metaphors, stark as they seem, have become real in London since the redevelopment of Docklands. As the operations of the State-appointed but not publicly accountable London Dockyard Development Corporation (set up in 1981) have been the subject of extensive analysis and comment, they have not been discussed here. Yet there is widespread consensus concerning the social outcome of the £700 million of public money invested by the development corporation run by property interests to attract both national and international capital in the redevelopment of redundant docklands. Once the centre of industrial working-class employment, whose local unemployment rate in 1987 had risen to 30 per cent, the docklands area, with little regard for the employment and housing needs of local people, has been ruthlessly redeveloped with new buildings and £200,000 luxury apartments, primarily aimed at wealthy buyers from the expanding financial sector of the adjacent City. For the 50,000 industrial jobs that have gone, only a fraction have been created.

Investment capital, from the Middle East, Japan, the United States, continental Europe, Latin America, and China (as well as from the UK) is drawn in to create new internationalized space, available to the highest bidder. As with other recent developments, there is a steady privatization of the public sphere. The new China Centre will provide 200 showrooms for 500 Chinese importers as well as a four-star hotel, a medical centre, apartment units, and a department store (*The Times*, 8 September 1986). In these, or other developments, old connections continue: the developers of 400 new luxury apartments (with yacht harbour) at Chelsea Harbour are the old, but restructured shipping conglomerate, the P and O Steam Navigation Company and the Global Investment Trust. The apartments are marketed by Savills in Hong Kong (*The Times*, 29 October 1986).

In Docklands, as house values tripled in two years, indigenous islanders are squeezed out while the Government 'applauds what it sees as the purest example yet of liberation capitalism, as practised in South Korea and Hong Kong' (*Time Out*, 19–26 November, 1986, p. 22). As the Far East is an increasingly lucrative market, Chesterton's, the estate agents, set up an office in Singapore and in 1985, 28 per cent of their prime London residential property went to the Pacific basin market (C. Whelan, 'Property', *Country Homes*, April, 1986). It is these and similar developments that have increasingly fuelled social and political conflict, which 'has been a central theme in docklands redevelopment', one of the largest areas for redevelopment in Western Europe (Page, 1987).

The emergence in the mid-1980s of the street-based 'Class War', engaged in vandalizing the expensive cars and property of 'yuppie' newcomers to Docklands (*The Observer*, 12 April, 1987), though hardly an 'urban social movement' is only one measure of the social conflict that these developments have generated.

More serious has been the growth in political confrontation, crime and especially, racial attacks, particularly against the Asian population in the East End. Unless major economic, social, and political changes are implemented in the near future, the problems of racism, unemployment, or social discrimination will lead to serious conflict in the city. While these comments need locating in a larger perspective and placed against other accounts,[11] Elliott (1986: 181) gives figures showing a rise of 73 per cent in recorded criminal offences in London between 1974 and 1984, and an increase in recorded crimes of violence of 98 per cent in the same period. Street robbery of personal property (mugging) increased by 400 per cent.

Many of the UK's estimated 50,000 narcotics (who, in 1984, spent an estimated £100 million on heroin) are in the capital. The increasing amount of money from drugs and crime generally has to be recycled

147

and much is going into 'some of the capital's most prestigious developments, including the Docklands' (Shawcross and Fletcher, 1987). In October, 1986, the Lord Chief Justice announced that 'a huge crime wave threatened to engulf Britain', which created a need for many more courts (*The Times*, 7 October 1987). Such Establishment views are used to justify the growing expenditure on policing and the increasing proportion of police in the London population. About a quarter of the country's entire police establishment is based in London (Elliott, 1986: 191) where there is one police officer to every 265 residents. This compares to police forces in areas with the next highest ratios such as Merseyside (1:324) and Greater Manchester (1:373), and where the national average is 1:458 (*Police Review*, 7 March 1986, p. 504). Apart from these two areas, expenditure by the Metropolitan Police Force per 1,000 people is roughly double that in other parts of the country. Between 1979–80 and 1985–6, spending in real terms on law and order in the UK increased by 27 per cent compared to a 4 per cent increase on education and a massive 68 per cent reduction on housing expenditure (*The Guardian*, 27 January 1985). In 1987–8, police expenditure was expected to rise by a further 13 per cent.[12] To these figures must be added the massive growth in private security services and equipment firms, the numbers of which rose, according to the Yellow Pages, from 10 in all London districts in 1960 to 245 in 1970 and some 850 in 1988.

In the capitalist world-economy, world-city roles (whether for London or other cities) bring major contradictions, not least between public and private sectors. Labour rewards in the upper echelons of the 'internationalized' private sector, determined by global criteria, drain skills from the national public sector.[13] The interests of international capital require public-sector spending on infrastructural facilities while simultaneously, they look for tax concessions to persuade them to stay. Reproduction costs, in labour, education, training, health, administration, law and order, are all borne by the State. And as in the colonial primate city, excessive investment in the world city leads to regional polarization and uneven development. These, and other issues, need further investigation in a comparative context.

Privatization of public housing plus the massive escalation of house prices fuelled by London's world-city role has increasingly meant that public-sector employees — nurses, teachers, council workers, and others — can no longer afford to live there. Hence, hospitals face chronic staff shortages and schools fail to find teachers. At the end of 1987, the Inner London Education Authority had 775 too few primary teachers; the cheapest one-bedroom flat in the capital cost five times the teacher's basic salary of £8,500 (*Sunday Times Magazine*, 3 January, 1988).

In the higher education sector, some of London's polytechnic institutions, initially established to serve the interests of the capital and also to provide both full- and part-time education for its population, were in the late 1980s being forced out of London by high land and transport costs, depriving the population of educational opportunities and the institutions, of access to centrally located facilities.

According to a strategic document of the late GLC's Economic Policy Group, 'Paid to think: professional workers in London' (1983), whilst the city had, in 1980, 17 per cent of national employment in all occupations, it had a much higher proportion in what they termed 'thinkwork' services. As percentages of the national total, London had 41 per cent of all employment in business services (including management consultancies, market research, public relations, etc.), 48 per cent of all in trade unions, professional, and business organizations, 28 per cent of all in professional and technical services (including architects, surveyors, consulting engineers, and supporting services), 26 per cent of all employment in libraries, museums, and art galleries, 14 per cent in higher education and 12 per cent in research and development (including scientific, medical, and social scientific R and D). As the GLC's strategy document points out, this concentration of people paid to do strategy and design work raises major questions with regard to workers who are *not* in these categories (not least, many women and ethnic minorities) in respect of the planning and effective control over their own work and the conditions in which it is done (Greater London Council, 1983a).

Attention has been drawn to the growth of Japanese and American banking and services in London and in particular sectors of education. But increasingly in the 1980s, health services too, including private hospitals and insurance have been seen as a potential growth area of American corporate interests.

The GLC's study of *Commercial Medicine in London* (Griffith, Payner, and Mohan, 1985) indicates the extent to which large commercial and mostly American hospital chains have bought British for-profit groups in recent years. The reputation of Harley Street, London's position as world city, its connection with medical markets in the Middle East and the high proportion of privately insured patients in the London area have been major attractions. (Over 10 per cent of the population in the Greater London area is said to be privately insured and between 1974 and 1984, the percentage of executive staff with private medical insurance went up from 30 to 69 per cent.) According to the report, the old 'charitable tone' of private practice is being increasingly undermined by strident commercialism as corporations take advantage of the Harley Street cachet (whose medical practitioners, by sharing addresses and mail boxes, have grown from 150 in 1900 to 1400 in 1985). The hospitals, buttressed by infrastructure support from the public sector and poaching nursing staff, are principally in the south and central areas.

In addition to examining London's banking and financial role, this study has also looked briefly at the growing importance of the cultural economy of London, the displacement of a trade in goods by a trade in signs, images, symbols, in the projected lifestyles of advertising, television, videos, and films. Culture, to a greater or lesser extent, is incorporated by capital. Yet in this respect, the world city also has its contradictions. At once a centre for the production and diffusion of a 'Western' mass culture, it is also, through the diversity of its peoples, its ethnicities, its sub-cultures, its alternative cosmopolitanisms, its representations of both core and periphery, also an instrument for changing that 'Western' culture and also, indeed, for changing the culture of the country wherein the world city is located. It is not only the economy which is being restructured but, also, the nature of national culture and identity.

As capital of the world art market, London has a particular importance, using art objects to store capital made elsewhere. In the first week of December 1986, for example, more money was spent on London's picture sales than ever previously recorded in auction history, some £83 million at Sotheby's and Christie's. The cause was the influx of capital from Wall Street speculations and others benefiting from takeover bids.

It is in this context that the purchase of Sotheby's by an American financier specializing in property development is understood. Between 1983 and 1987, by imposing American corporate-management techniques and new financial disciplines, the value of the company was quadrupled with turnover in 1986-7 up 77 per cent on the previous year. Of the 7.2 million shares, however, 63 per cent were offered for sale in the United States, 20 per cent in Britain, and 16 per cent on the international art market (*The Independent*, 6 October 1987).[14]

Fine art is a movable commodity; residential property is not. Yet the UK's particular concern with 'period' houses is also exploited on the international scene. One particular estate agent (Knight, Frank, and Rutley) looks on houses 'as a piece of art' according to *Property International* (1 March 1985); 'they are to the property world what Sotheby's are to the world of antiques'. Like other firms in the 1980s, their 'period houses' have largely been used for national and international investment.

These developments, which are tied to the larger growth of London as world city, draw on its clientele of international elites. Like banking, securities dealing, education, or culture, London is used as a *site*, in the case of medicine, as a recruiting ground for employing doctors for the Gulf, or, in business and management services, as a place to headhunt corporate executives. In the present political environment, such developments undermine public-sector facilities both by poaching labour and eroding infrastructural investment.

The strength of international corporate interests and the internationalization of the elite impacts on government and other decision-makers alike. As in the colonial city or the Treaty Ports of China, the international elite have temporary commitments to where they stay; their interest is in business, their moral attitudes 'disengaged' (Townsend, 1987: 70). Their influence is also extended to the local courts.[15]

But whether outside the State or within it, their interests bear heavily on government policies. (Even the Thatcher government's decision to ban dealings with Argentine banks during the 1983 conflict was over-ruled by the major banks acting as part of the international financial system (CIS, 1983)).

A brief comment is needed on the built environment. Conventional wisdom in the social sciences has for long maintained that environments, built or natural, cannot be 'determining'; they do not, in short, make significant differences to economic and social change. The result has been that in the analysis of recent urban and regional restructuring discussion of built form has been omitted.

Yet it is evident from events in recent years that the absence or presence of particular environments has been a major influence affecting, as also necessary for, economic and urban restructuring. The massive investment in office space in the last few years will, by

151

providing such quantities at the expense of other infrastructural developments, both in London and elsewhere, powerfully influence if not preempt, the shape of future employment in London. And it is here that London's dependence on the volatility of global markets is at its greatest.

By the end of 1988, the effects of the Stock Market crash were clear. According to the *Chartered Surveyor* (15 December, 1988), 'the party in the City office market is well and truly over' (the same could have been said about the rise in house prices). There had been 'tremendous over-capacity in the securities market' and after the crash, according to *The Banker* (November, 1988), fifteen banks and security houses had left. Though the total of foreign banks remained at 448, employment in them had dropped back by 20 per cent to 58,000. The reduction in trading volume in securities had resulted in redundancy for 5000 dealers.

In the face of these developments, in early 1989 the office market was much more sober. Many of the developments specially designed for 'Big Bang' were either unlet or not being used for the purposes for which they were intended. Yet the Canary Wharf and other developments on the Isle of Dogs have gone ahead, creating a further 18 million square feet of office space, equivalent to 25 per cent of the entire floor space in the City. Though there was a real threat of over-supply and property developers were far more pessimistic than a year before, they were looking to 1992 and the critical imperative of maintaining London's role as the financial centre of the single European market.

Finally, some remarks are needed on the world-city approach adopted in this study. Whilst the theoretical propositions used as guide lines for this study provide a set of valuable tools, a number of riders are needed. As Korff (1987) suggests, world cities in Europe are far more integrated with their larger 'national' economies, even with the European space economy in general; in terms of world-city functions and concomitant characteristics, Friedmann's 'urban region' would need to take account of 'not only the whole of the South East (and) much of the neighbouring two regions' (Hall, 1987: 105), but, in regard to particular functions and effects (not least international property investment and the production and export of producer services), a much larger part of the British and even European space economy.[16] London is, in many ways, a metaphor for the whole country, its range of inequality reflected in the North–South divide. But this is not to undermine the relevance or value of the world-city approach: as we have seen, London and its region concentrates most of the functions and characteristic attributes in itself; they are not, however, limited to it.

Another aspect of the world city hypothesis is that it perhaps tends to focus too much attention on the pathologies of the world city (not least in this account), on its negative rather than positive attributes and potentialities. And by focusing primarily on its economic aspects (as has also been the case here), it neglects the cultural and especially the political aspects which they fuel and inform. In some ways, the world city is a microcosm of the world itself, though it is clear from this account that each world city has its own very distinctive history, its own very distinctive social and cultural composition and its own distinctive location within a particular culture and state, linked though it may be to a web of regional and global connections. With its mix of cultures, not least of peoples once marginalized on the global periphery, within a democratic state, it then presents an opportunity for forging a new consensus, an opportunity for addressing the pressing issues of the contemporary world be they political or cultural, ecological or economic.

Little mention has been made in the earlier sections of London's role in Europe. Hesitating for long before joining the European Community, and reluctant to abandon her old international role, the UK fell between two stools in the 1960s and was politically ambivalent about 'Europe' in the early 1980s. In terms of hosting 'European Community' institutions, compared to Paris, Brussels, or Luxemburg, London's status is defined more by its international and post-imperial history than its historical connections with the European continent. Few Community institutions are in London or its region (the European Centre for Medium Range Weatherforecasting, the European Development Fund Information Service, the European School, near Oxford, and possibly, the European Patents Office in the future). 'European' power, as its political decision-making, is elsewhere. What London has provided is, again, the site, the articulation point between continents, manifested, for example, in the Euro-Arab bank, the Euro-Latin American Bank, the base for Japanese and American multinationals setting up operations in Europe, or the consumption (rather than production) site for European Community goods. And whilst London belongs to a European urban system (Cheshire and Hay, 1986), as the previous pages have shown, it equally belongs to a world, and particularly, a post-imperial world system that equally influences its role in Europe. The effects of the full 'economic integration' of Britain with the European Community in 1992 will have massive long-term implications for London, though this question, a significant problem for research, is beyond the scope of this discussion, not least the immense effects of the Channel Tunnel.

London (with the South-east) is a bridgehead. Its growth is fostered by national and international pressures coming from four sides. From (and to) the North, there is domestic migration; from the West, influences from across the Atlantic; from the South and Europe, across (and soon under) the Channel; and from the East, by air, and from the Pacific Basin Japan and Hong Kong.

The themes discussed in this study do not pretend to present the entire picture but rather, a particular dimension of it. As Walton (1982: 125) has written concerning urbanization on the periphery of the world-economy 'the global economy is one among several powerful influences . . . an influence that has direct, indirect and remote effects, all of which interact with concrete local influences, including social organization and cultural tradition'. Yet because of Britain's (and London's) past history, present institutions, linguistic and cultural characteristics, and (not least) recent particular political orientation, London is far more dependent on the world economy than many other cities in the world (see also Thrift, 1985) and hence, is consequently more vulnerable to changes in it. There is no guarantee that particular countries or cities will, in the long historical term, remain either at the core or the periphery. And a major collapse in the world-economy will hit harder those cities whose fortunes are most tightly tied to it than others which are not, as has recently been seen.

The conclusion from this is also that reached by Grunwald and Flamm (1985): a massive investment is needed in upgrading the workforce, which means huge investments in education, training, and alternative policies for employment and development (see also GLC, 1985 and Townsend, 1987). Yet what is planned by the government for London are 50,000 new jobs in tourism by the 1990s with priorities placed on 'removing homeless families from bed and breakfast and cheap hotel rooms and increasing the number of beat policemen to improve safety'. The strategy of the London Tourist Board with extra resources from government, aims to have 12 million overseas tourists a year (plus 15 million from other parts of the UK) by the early 1990s (*The Times*, 30 June 1987).

There is no doubt that, despite their very significant differences, specific world cities in the capitalist world-economy have structural features in common that are identified by world-city literature. The extent to which they differ is explained not only by social and cultural factors but also by state policies: thus political decisions by Japan have inhibited extensive international labour migration. In other respects, Rimmer sees the world city perspective as valid, though seeming 'to downplay the importance of internal forces' (Rimmer, 1986: 152).

The most important cultural feature that distinguishes Tokyo and other world cities from London and New York but unites certain others (specifically, ex-British colonial ones) round the world is that of language, a critical dimension which to date, has received inadequate attention in research. With other variables, it is this (as Dunning and Norman (1987) suggest in another context) which is critical to London's continuing role in the world-economy. Yet where language is frequently seen as a fortuitous advantage for London (and the UK) in competing in the world-economy it acts equally to her disadvantage. Linguistic space, like the territorial space of London itself, is simply a convenient terrain on which international capital operates. Without economic and political power to back it, the linguistic imperialism of English is no guarantee of success.

Given the long historical role of London in the world-economy and its privileged language and cultural setting, the critical question is the extent to which recent and future developments have been determined by the role of the State. Whilst discussion of the relationship between State and capital is neither possible nor appropriate at this point, it is clear that recent political decisions have been of major significance but also, that larger structural interests have been equally important in influencing these political events. And it is by focusing on an understanding of these structural features that political and social alternatives become possible.

What world-city approaches make clear is the necessity not only to look at the world beyond the State but to look at the world as one. The case for research beyond national boundaries is as strong, if not stronger than for looking at phenomena within them. The understanding of one world city requires the understanding, in a comparative context, of others, but also the understanding of itself within a long-term historical perspective. The aim of this study has been to provide for London some benchmarks from which its future role and development in the world-economy can be charted. And by arguing for a global perspective, to provide insights into contemporary developments. Understanding opens up the opportunity for action.

APPENDIX 1

SUGGESTED CHARACTERISTICS OF COLONIAL CITIES

(See King (1989b) *Urbanism, Colonialism, and the World-Economy*, Chapter 1

Geo-political

1. External origins and orientation.

Functional

2. Centre of colonial administration.*
3. Multiplicity of functions, with presence of banks, agency houses, insurance companies, etc.
4. Focus of communications network.
5. Acts as economic intermediary — symbolized by corrugated iron *godowns*.

Political/economic

6. Dualistic economy, dominated by non-indigenes.
7. Presence of large group of indigenous unskilled and semi-skilled migrant workers (which see colonial city as alien community).
8. Municipal spending distorted in favour of colonial elite.
9. Dominance of tertiary sector.
10. Parasitic relations with indigenous rural sector.

Political

11. Eventual formation of indigenous bureaucratic–nationalist elite.
12. Indirect rule through leaders of various communities.

Social/cultural

13. Social polarity between superordinate expatriates and subordinate indigenes.
14. Caste-like nature of urban society.
15. Heterogeneous dual, or plural society with three major components:

    (a) Elite formed by residents of colonial/imperial power with externally derived authority based on military force;
    (b) Intervening groups that originate from racial mixing and in-migration from other colonial or semi-colonial territories (e.g. overseas Chinese);
    (c) In-migrated indigenous resident groups consisting of educated modern intelligentsia and modernizing elites, as well as uneducated ethnic groups, clans, etc.

16. Occupational stratification by ethnic groups.
17. Pluralistic institutional structure.
18. Residential segregation by race.
19. Large groups of unskilled indigenous and semi-skilled migrant labour.

Racial/ethnic

20. Racial mixing.
21. Occupational stratification by ethnic groups.
22. Racial residential segregation.

Physical/spatial

23. Coastal or riverine site.
24. Establishment at site of existing settlement.
25. Gridiron pattern of town planning combined with racial segregation.
26. Urban form dictated by 'Western' models of urban design.
27. Specific character of residential areas.
28. Residential segregation between exogenous elite and indigenous inhabitants.
29. Large difference in population densities between areas of colonial elite and indigenous population, impacting life style and quality of life.

30. Tripartite division between indigenous city, civil, and military zone.

*some exception to this

APPENDIX 2

BRANCHES OF FOREIGN BANKS FUNCTIONING
IN INDIA, 1987

The number and location of branches of foreign banks in India provide
a useful insight into, amongst other facts, both the Indian diaspora
round the world, lending countries funnelling investment into India,
and the capital flows (including remittances to India) between different
countries and various locations in India. The overall predominance
of UK bank branches (71 out of the total of 136) and their Indian location is worth noting.

The following lists the principal foreign banks with three or more
branches.

Table A1 Foreign banks in India, 1987

Bank	Country of incorporation	Places where branches functioned and number		Total number of branches in India
1. Grindlay's Bank PLC	UK	Calcutta	(18)	56
		Bombay	(12)	
		New Delhi	(10)	
		Madras	(4)	
		Amritsar	(2)	
		Cochin	(2)	
		Bangalore		
		Darjeeling		
		Kanpur		
		Tuticorin		
		Simla		
		Srinagar		
		Gauhati		
		Hyderabad		

Bank	*Country of incorporation*	*Places where branches functioned and number*		*Total number of branches in India*
2. Standard Chartered Bank	UK	Calcutta Bombay Madras Delhi Kanpur Amritsar Cochin Goa Calicut	(8) (6) (3) (2)	24
3. Hong Kong and Shanghai Banking Corporation	Hong Kong	Calcutta Bombay New Delhi Madras Visakhapatam	(9) (7) (2)	20
4. Banque Nationale de Paris	France	Bombay Calcutta New Delhi	(2) (2)	5
5. Bank of America	USA	Bombay Calcutta New Delhi Madras		4
6. Algemeine Bank Nederland N.V.	Netherlands	Bombay Calcutta	(2)	4
7. American Express	USA	Bombay Calcutta New Delhi		
8. Bank of Tokyo	Japan	Bombay Calcutta New Delhi		3

Source: International Banking Section, Ministry of Finance, Government of India, New Delhi. I am grateful to Mr S.R. Gupta of the Indian Council for Research on International Economic Relations for this information.

Note: In 1987, 21 foreign banks had 136 branches in India: the UK, the USA, and France had 3 banks each; Japan and the United Arab Emirates had 2 each, and 1 each from Hong Kong, the Netherlands, West Germany, Bangladesh, Canada, Bahrain, Oman, and the Cayman Islands. Nineteen of the 21 had their principal office in Bombay, the other 2 in Calcutta. The largest number of branch offices were operating in Bombay (45), Calcutta (44), Delhi (19), Madras (10), Cochin (3), and Kanpur (2).

NOTES

CHAPTER ONE

1. In the following analysis, the term 'world-economy' is used in the
   sense as described by Wallerstein:

   > The concept 'world-economy' . . . should be distinguished from
   > that of 'world economy' . . . or international economy. The latter
   > concept presumes there are a series of separate 'economies' which
   > are 'national' in scope, and that under certain circumstances these
   > 'national economies' trade with each other, the sum of these
   > (limited) contacts being called the international economy. . . . By
   > contrast, the concept 'world-economy' assumes that there exists an
   > 'economy' wherever (and if but only if) there is an ongoing exten-
   > sive and relatively complete social division of labor with an
   > integrated set of production processes which relate to each other
   > through a 'market' which had been 'instituted' or 'created' in
   > some complex way. Using such a concept, the world-economy is
   > not new in the twentieth century nor is it a coming together of
   > 'national economies', none of the latter constituting complete divi-
   > sions of labour. Rather, a world-economy, capitalist in form, has
   > been in existence in at least part of the globe since the sixteenth
   > century. Today, the entire globe is operating within the
   > framework of this singular social division of labor we are calling
   > the capitalist world-economy.
   >
   > (Wallerstein, 1984: 13)

2. See, for example, the following opening paragraphs from the *Yorkshire
   and Humberside Development Review* 1, 7, June/July 1988, 'City takes on
   a new image' (p. 12):

   > A bold, ornate, rich terracotta building spans one side and
   > dominates Park Square in the centre of Leeds. The extravagant
   > Moorish facade of the old mill, with its columns and cresting, is
   > capped with pinnacles. It shrieks of Victorian success and bygone
   > splendour.

But there is nothing faded about the massive building — it has been carefully, and expensively, restored and refurbished. And it seems to epitomise the switch in the city's economic base and development in recent years.

St Paul's House was built as a prestigious factory in 1878 by John Barran, the pioneer of mass-produced clothing — one of the principal industries upon which the wealth of the city was built in the last century and the early part of the present one.

It is now owned by the Norwich Union Pensions Management, who bought it for £4 million. Not far away, the facade of another former clothing factory fronts a new investment by Legal and General. These are just two such developments which reflect the fact that in the rebuilding of the Leeds economy after its semi-demolition in the Seventies and early Eighties recession, Mills have moved over to make way for Money.

The grand city which was literally built on the thriving textile and engineering industries of the Victorian era (and at least half the buildings in the centre are protected) is steadily taking on a new image.

Increasingly, the city is being referred to as the business centre not only of the region but of the North.

. . .

In the business services sector, the figures would appear to show that Leeds has completely outstripped the rest of Yorkshire and Humberside with a growth rate of 15 per cent compared to 7.2 per cent in such jobs as insurance brokers and agents, solicitors, accountants, auditors, architects, estate agents and office staff. (p. 12)

According to a research officer with Leeds City Council's Department of Industry and Estates, the concentration of finance and business services in Leeds is increasing and companies were moving to the city and expanding their current businesses there. 'The four main clearing banks have their regional offices in the city, in which a total of 27 UK, international and foreign banks are now represented. It is also the headquarters of the Yorkshire Bank. Several major insurance companies have moved a considerable amount of their business to Leeds' (including the Sun Alliance, Norwich Union, Guardian Royal Exchange, and seven large building societies, such as the Leeds and Holbeck and the Leeds Permanent, have major offices in the City).

The article also refers to:

The Corn Exchange — an historic listed building that is being refurbished as an imaginative specialist shopping centre, . . . a converted warehouse on the refurbished riverfront has also provided a home for the UK's first ever Design Innovation Centre. . . the fine Victorian shopping arcades where the best of

162

the past is being incorporated into the new shopping developments currently taking place.

The new 'mini-Urban Development Corporation' has led to the creation of a public/private partnership backed by P and O, 'one of the leading international construction and development groups and Mountleigh, a Leeds-based international property group' (pp. 11-13).

In the ten years to 1988, employment in manufacturing in Leeds has declined from 700,000 to 440,00; in financial services, the number has risen in the same period from 68,000 to 149,000 (of whom two-thirds are women, mainly in the less-well-paid occupations). One person in twelve worked in the financial services industry in 1988. Between 1983 and 1986, income generated in financial services jobs increased from £2 billion to £3.1 billion and unemployment levels have fallen from 15 to 10 per cent. These developments have had a major impact on employment structure and also, on the price of both commercial and residential property, the latter increasing by 30 per cent between April and September, 1988 (information from Ken Woollmer, Department of Business and Economic Studies, University of Leeds, December 1988). It is against this background that the City Council opened negotiations in 1988 with the Canadian developers, Triple Five Corporation, responsible for the massive West Edmonton Mall in Canada, with a view to their building a £3 billion shopping, leisure, and housing complex on a 100-acre site south of the Leeds city centre, stretching from Boar Lane across the river Aire into Holbeck. According to City councillors, 'It is a major world investment unique to the UK and it would virtually wipe out unemployment in the city. . . . It will make Leeds a tourist centre and attract millions of visitors' (*Yorkshire Evening Post*, 21 December 1988, pp. 1, 3). (It is equally likely that, in future, cities *without* such complexes will have greater tourist appeal.)

CHAPTER THREE

1. The significance of the role of the state in transforming once-colonial cities is, of course, crucial. Burma, with a highly centralized, socialist state, has kept out foreign capital and in 1987 had only one joint venture in operation. The capital, Rangoon, remains as a classical example of the archaeology of colonialism with relatively few modifications made to the built environment in four decades.

    Karachi (Pakistan) or Colombo (Sri Lanka) present a completely different face. Whilst both were strongly influenced by colonialism and manifested many of the archetypical characteristics of the British colonial city in 1947, in the last fifteen years, both have undergone major transformations in the context of liberal, market-oriented state policies. As in Delhi, there are high-rise offices, international banks, multinational headquarters buildings, and luxury hotels. As elsewhere in world cities, economic and social polarization has been rife. As an example of the impact of Western capital, the salary of the secretary to the Chief Executive of American Express in Colombo was reported to be higher than the Chief Permanent Secretary in the government's Ministry of Finance (personal communication, Page, 1987; for an account of 'Town planning and the neo-colonial modernization of Colombo' see Steinberg (1984)).

CHAPTER FOUR

1. The distinction between colonialism and imperialism can, as Drakakis-Smith (1987: 11) implies, 'lead to a tortuous debate'. Imperialism, associated with the phase of monopoly capitalism and involving the military and political expansion of nations beyond their borders, largely for economic purposes, tends to focus attention on the imperial powers at the core; colonialism, whether of an exploitative or settler variety, on the colonies at the periphery. They are different sides of the same coin. See also King (1989a).

   It is worth noting that, at its demise, the British empire was formally referred to as 'The British Colonial Empire'. In London, both imperial and colonial functions could be identified, dependent on the economic, political, social, and hence, constitutional status of the states referred to.

2. It has been suggested that, in the later nineteenth century, British manufacturers sought to expand in the less-developed areas of the world (such as Africa and South-East Asia) because they were unable to compete with more advanced countries in Europe and North America (Hopkins, 1980).

   In the new advanced producer sections of the information society of the late-twentieth century, in the world market for education, the same process is being re-enacted; northern universities, with their origins in nineteenth-century industry, faced with declining resources, collaborate to recruit fee-paying students not in the fiercely competitive global market of advanced post-industrial nations of North America or the European Community but exploit historical and cultural connections and concentrate on the old colonial markets of less-developed societies (see P. Radcliffe (1986), *Take 5. Graduate Studies at Five British Universities*, University of Manchester). The Universities of Birmingham, Leeds, Liverpool, Manchester, and Sheffield finance a joint campaign aimed at 'overseas students', a term apparently equated with 'students from developing societies', viz: 'many of the courses offered have been developed specifically to meet the needs of overseas students — courses on health and education, administration and development, planning and economics'. Mention is made of 'particularly large groups of students from Malaysia, Hong Kong, Nigeria, and the Middle East'. The long-term consequences should be clear.

3.   There were those who would not admit it — even perhaps some here today — people who have strenuously denied the suggestion but — in their heart of hearts — they too had their secret fears that it was true: that Britain was no longer the nation that had built an Empire and ruled a quarter of the world. Well they were wrong.

(quoted in Cowan, 1987: 15)

The colonial (and military) influence on Cheltenham also extends to the nomenclature of the built environment. Local Council housing blocks built in the 1950s and 1960s along Princess Elizabeth Way are called Rhodesia House, Ceylon House, Australia House, and South Africa House; other names are Hobart, Tasmania, Quebec, Canada, Montreal, New Zealand, Durban, and Auckland.

Elsewhere, there is Imperial Square, Imperial Lane, Imperial Circus, Imjin Road, Burma Road, Somme Road, and Ladysmith Road, the latter also suitably close to the Government Communications Head Quarters. (See also King, 1976, 'Urban nomenclature', pp. 246–8.)

I am grateful to Harry Cowan, Gloucester College of Arts and Technology, for this information.

CHAPTER FIVE

1. Tables 5.11(a) and (b) reproduced from Scott (1867) provide a useful insight into the continuities and change in the preferred residential locations of City workers.
2. Whilst telephone directories provide a valuable source of data (not least in making comparisons between cities) their use clearly presents problems of classification that are not underestimated or discussed here, for example, in distinguishing between the significance of the use of 'International' in the 'International Airline Passengers Association' and 'International Stores', or between 'Imperial Chemical Industries' (recently changed to 'International') and 'Imperial', the name of a pub or hotel): A more detailed analysis of these data is planned, along with other more formal sources on international organizations.
3. Examples of internationally organized activities, whether primarily economic, commercial, political, cultural, ideological, or administrative in purpose, though in many cases including funding from outside the state, include:
    International: Accounting Standards Committee, Advertising Association, Alliance of Women, Bank for Investment and Commerce, Bank for Reconstruction and Development, Bar Association, Bullion and Metal Brokers, Celebrities Management, Cerebral Palsy Society, Cocoa Organization, Colour Authority, Commodities Clearing House, Computer Programs, Copyright Bureau, Defence Aid for South Africa, Dental Federation, Equipment and Materials

Table 5.11(a) City men of *six* specified commercial classes, enumerated as sleeping inhabitants of the several Metropolitan districts (1861)

Districts of Metropolitan Board of Works.	Merchants.	Bankers.	Stock and Commercial Brokers.	Ship Owners and Brokers.	Accountants.	Commercial Clerks.	Totals.
CITY OF LONDON..........	356	9	33	54	59	773	1,284
St. George's, Hanover Sq. ...	197	33	44	18	43	373	708
Marylebone	345	23	101	22	60	613	1,164
St. Pancras	306	14	114	61	144	1,010	1,649
Paddington	Given in the census with Kensington.						
Islington	563	27	211	110	216	2,039	3,166
Lambeth	323	7	170	76	111	1,274	1,961
Kensington	707	47	228	49	100	761	1,892
St. James's, Westminster	134	7	17	9	3	168	338
Lewisham	222	6	109	51	49	316	744
Hackney	277	13	166	94	149	1,465	2,164
Wandsworth	194	15	87	19	44	277	636
Poplar	22	0	20	60	20	245	367
Westminster	46	6	21	3	30	234	340
Chelsea	40	6	35	7	31	222	341
Strand	74	6	12	5	24	201	322
Shoreditch	57	0	26	18	73	739	913
Whitechapel	50	0	13	14	15	140	232
Greenwich	85	1	56	65	53	317	577
St. Gile's	136	13	33	7	24	322	535
St. Martin's	62	3	5	7	5	123	205
Camberwell	197	5	107	58	85	858	1,310
Clerkenwell	57	0	21	10	31	369	488
Newington	87	8	32	23	68	772	990
Limehouse	13	0	3	32	7	113	168
St. George-in-the-East	6	0	3	18	6	81	114
Holborn	52	1	27	11	27	299	417
Rotherhithe, etc.	14	0	5	11	5	138	173
Bethnal Green	14	1	20	4	21	295	355
Mild End Old Town	25	0	17	41	38	504	625
St. Luke's	71	1	11	4	22	145	254
Fulham	Given in the census with Kensington.						
St. Saviour	14	0	7	1	8	76	106
Bermondsey	13	0	9	11	20	188	241
Hampstead	132	11	37	21	17	101	319
St. George, Southwark	30	0	8	6	13	180	237
Woolwich	Given in the census with Greenwich.						
Totals	4,921	263	1,808	1,000	1,622	15,731	25,345

Source: Scott, 1867: 26–7.

Table 5.11(b) Merchants, etc., sleeping *beyond* the Metropolitan Districts. [Census of 1861.]

District	Merchants.	Stock and Commercial Brokers.	Accountants.	Commercial Clerks.	TOTALS.	District	Merchants.	Stock and Commercial Brokers.	Accountants.	Commercial Clerks.	TOTALS.
Barnet	16	11	...	32	59	Kingston	53	48	11	61	173
Brentford	52	25	27	90	194	Reading	16	7	13	33	69
Brighton	74	21	44	99	238	Reigate	36	19	6	26	87
Bromley	49	18	4	30	101	Richmond	39	21	13	45	118
Chertsey	21	10	3	5	39	Romford	27	10	5	37	79
Croydon	155	106	25	169	455	Staines	10	7	5	10	32
Dartford	26	26	12	32	96	Steyning	17	7	4	16	44
Edmonton	89	114	32	275	510	Tunbridge	13	7	8	25	53
Eltham	18	5	7	16	46	Uxbridge	9	1	5	29	44
Epping	9	14	2	14	39	Ware	10	2	1	15	28
Epsom	36	21	7	34	98	West Ham	97	78	29	241	445
Eton	10	2	3	19	34	Worthing	11	4	—	13	28
Guildford	13	2	4	32	51	Windsor	15	11	3	23	52
Gravesend	10	22	9	39	80						
Hendon	27	8	6	34	75	Totals	958	627	288	1,494	3,367

Source: Scott, 1867.

for the Arab World, Exhibition and Conference Services, Federation of Actors, Federation of Business and Professional Women, Financial Markets, Gold Corporation, Hospital Equipment, Importers and Exporters, Institution for the Conservation of Historic and Artistic Works, Medical Personnel, Mergers and Investment Service, Military Services, Missions, Muslim Movement, Oil Insurers, Planned Parenthood Federation, Publishing Corporation, Publicity, Risk Management Services, Ship Brokers, Society for the Protection of Animals, Sugar Organization, Talent Bookings, Tennis Federation, Timber Corporation, Tin Council, Trade Development, Transport Workers Federation, Union of Aviation Insurers, Villas, Wheat Council, Wool Council, and Year of Shelter for the Homeless.

4. In 1986, 84 per cent of the top 50 British companies had their headquarters in London and 75 per cent of the top 100. Of all the 500 biggest companies, 75 per cent have their headquarters in London or the south-east of England (*Business*, October 1986).

In the early 1980s, London company headquarters included, in Victoria and Westminster: British American Tobacco, Bass Charrington, Bowater, Dunlop Holdings, Imperial Group, and Rio Tinto Zinc; in the W1 district: EMI, Courtauld's, General Electric, Glaxo, and Reed International. In the City: British Petroleum, Allied Breweries, Brook Bond, Consolidated Gold Fields, Tate and Lyle, and Unilever; in the central area: GKN, BICC, Cadbury Schweppes, and Thorn Electrical. The names of some of these companies occupy most of the first fifteen places of the UK's largest industrial companies: of these, four are in oil, three in tobacco (plus retailing, paper, and food — though increasingly in printing and publishing), two in motor vehicles, two in hotels, and one each in chemicals, industrial, and manufacturing. Significantly, in the last two years, two advertising agencies have figured in the list of the top ten *The Times 1000*, 1986–7). About thirty of Britain's largest 500 companies are in paper, printing, and publishing.

5. See particularly Landsdowne College, Schiller International University, Webster College, The American College of London, the American College in London, Richmond College, etc.

6. Those listed in the London directories' Yellow Pages include Ball State University; Brigham Young University; Drew University; Findley College, Ohio; Pepperdine University; Samford University, Alabama; Wake Forest; Grenell College, Iowa; University of Southern California: University of California; Regents College (Rockford College); and others. Other universities and colleges hire premises for use in London (see also the American Institute for Foreign Study). The American University of London is apparently a correspondence college.

Other new educational institutions have resulted from the growth, particularly since 1970, of large cohorts of foreign nationals in the city, who have set up schools in different parts of London. First established

168

in temporary premises, these have later given rise to purpose-built institutions as the populations for whom they cater become more permanent. One of the oldest, the Lycée Français in Cromwell Road, South Kensington (founded in 1919) concentrates many French-speaking residents in that area, and others on connecting underground lines to Wimbledon, Richmond, Kew and Ealing. The Japanese School, with 800 daily pupils and another 1,600 usually in English schools but enrolled for Japanese language classes at the weekend, established in the early 1970s, recently moved from Camden Town to Acton, taking many of the resident Japanese community with it. The American School, with 1,180 students, started in 1952 on three sites, moved into purpose-built quarters in St John's Wood in 1971. In 1988, the two main American catchment areas were NW8 and NW3 (Hampstead), as well as sizeable groups in West Hampstead, NW6, Notting Hill Gate, Holland Park, SW3, SW1 and SW7. The German School, started in 1972, moved into purpose-built premises in Richmond in 1981. Its main catchment areas are in Richmond, Wimbledon, Kingston upon Thames, Kensington and Knightsbridge. More recently, the King Fahad Academy in Acton has catered for pupils from the Middle East.

As housing for these residents is often financed by foreign employing companies or governments, these developments have a major impact on local property prices and over the years, have significantly increased the foreign ownership of domestic property in London, for investment, for 'holiday houses', and temporary use, or, to subsequently let to the owner's own company or government (Doreen King, 'The foreigners' favourites', *The Times* (Property Section), October 29, 1988, p. 12).

7. A representative engineering firm, established some thirty years ago, has its headquarters office in W1 and employs a staff of 2,600 of whom 1,100 are based in London, 1,100 overseas, and the remainder elsewhere. A multidisciplinary consulting engineering service is provided in civil, mechanical, electrical, building, geotechnical, structural, and industrial engineering: overseas offices are maintained in France, Hong Kong, Kuwait, Libya, Qatar, Saudia Arabia, and, with partners, in Brunei, Ghana, Liberia, Malaysia, Mauritius, Papua New Guinea, Singapore, South Africa, Namibia, Zambia, and Zimbabwe.

Representative of architecture is a firm with offices in W1 that has completed factories and offices for multinational companies in Nigeria, university buildings, hospitals, and technical institutes in Ghana, a teaching hospital and medical-faculty building in Malaysia, a theatre in Singapore, an embassy in Indonesia, buildings for the University of Benghazi and of Tripoli, the Benghazi city-centre development plan, etc.; other firms specialize in different areas and building types: an international airport at Sharjah, broadcasting studios at Oman, a cultural centre and theatre in Jordan, a villa estate for the rule of Dubai, etc.

In design consultancy, a representative medium-to-large firm is located in W1 and employs 150 professional staff with other offices in Paris, Madrid, Dubai, Abu Dhabi, and Johannesburg, servicing a wide range of clients 'from small independents to large multi-nationals'. Set up some twenty years ago, it specializes in retail design, office planning, and design, travel, and transit design, graphic communication, hotel services, and project management. Turnover in recent years is some £4 million a year; clients include American Express, Middle-Eastern Airlines, and oil companies.

8. The Design and Art Directors' Association (with 500 members), the Society of Industrial Artists and Designers, the International Graphical Designers Association, the Design Council, and the Royal Society of Arts perform some of the professional functions.

9. As some architectural firms are listed both in the Central London as well as North, the North-west or the South-west directories, the figures indicated are likely to overstate firm numbers by 10–20 per cent. The absence of entries for 'design consultants', 'designers', and 'interior designers' in the directories published in 1960 may partly be accounted for by the listing of some of these categories under 'artists' or commercial artists' at that date. Only the category of 'industrial design' is given in 1960. The table is included principally as a stimulus to further research.

10. Though not always. On 7 January 1986, *The Times* carried a report: 'A penniless man (from Nigeria) wandering the streets of London's West End stabbed a wealthy American banker to death because he was jealous'. The assailant was alleged to have told detectives that the act was committed because the victim, an executive for Citibank, was dressed well and 'looked as if he went to all the posh places'. According to *The Times* report, counsel for the prosecution said that 'the two men were total strangers and fate brought them together in Albermarle Street, Piccadilly'.

11. The Greater London Council's *London Industrial Strategy* (GLC, 1985) provides the most comprehensive coverage of economic and selected social aspects of London. Elliott's (1986) readable account is strong on statistical information and contains useful accounts of housing, popular culture, and, critically, of the Metropolitan Police. It does, however, treat London as separate from both the rest of the world and from the UK.

12. I am grateful to the Police Monitoring and Research Group of the London Strategic Policy Unit for this information. (The Unit has subsequently been closed with the abolition of the Greater London Council.)

13. In 1986, according to the *Civil Service Commission Annual Report*, up to 40 per cent vacancies in particular civil service categories requiring computing, accountancy, science, actuarial, and legal graduate skills were noted.

14. These developments must be seen in relation to the use of 'art' objects as vehicles for the storage and circulation of capital as well as the use of 'art galleries' for urban regeneration and symbols of corporate power. Though still a long way from matching New York, with over 800 art galleries, greater Los Angeles had 150 in 1988, a roughly fivefold increase in the previous 20 years.

> Having replaced San Francisco as the West Coast's financial center in the 1980s, Los Angeles has spawned a monied elite who seek to affirm their culture as well as their clout by giving gifts to museums [though] generous donations, like designer clothes, tend to come with personal labels attached to them. Last year, banners outside the Los Angeles Center County Museum of Art touted 'the Wolper Picassos', a show of works from the collection of film producer, David Wolper.
>   (Michael Small, 'Thanks to Medicis like Norton Simon and Armand Hammer, a gilded lily of an art scene flowers in L.A.',
>   *People*, 28 March 1988, pp. 87–93)

In this transatlantic investment bonanza, the marketing of meaning has assumed major proportions, giving a filip to a particular kind of cultural production, not least in the social dimensions of 'architectural history'. Consider, for example, the following copy in a brochure from Sotheby's International Realty office in Madison Avenue, New York, for a house being offered for sale in Grosvenor Square, London:

> Situated on the north east corner of Grosvenor Square. . . in the heart of London's Mayfair, (this) is the last remaining building suitable as a private residence.
> This imposing and elegant Georgian house dates back to 1729 and was built for James, Earl of Northampton at the time the Grosvenor Estate was developed.
> The property occupies an unrivalled position midway between the United States Embassy . . . and Claridges Hotel in Brook Street and overlooks Sir William Reid Dick's statue of President Roosevelt which is at the centre of the 8 acre square . . . it is within a few minutes walk of Bond Street, Berkeley Square, Park Lane and Hyde Park and within a short drive or underground journey from the financial and business centres in the City.
> The square . . . houses the Canadian High Commission, the Indonesian Embassy . . . The house has previously been inhabited by the Hon. George Townshend (later to become 1st Marquis Townshend), Lord Bolingbroke, Lord Amherst, John Quincy Adam, the first American Ambassador to Britain, and Lord Cunliffe, governor of the Bank of England. The property has recently been occupied by an international bank. . . . It

would be ideally suited for occupation as an important private residence, for diplomatic use, to house an art collection or the corporate headquarters of a major corporation.

(Sotheby's New York, March 1988)

15. A London magistrate called yesterday for a relaxation of the immigration laws to allow foreign workers to take Britain's 'unwanted' low paid jobs. 'It is very difficult to see what public mischief is being done when people come from another country and take employment at competitive wages, work which nobody else wants. (The stipendiary magistrate was hearing the case of a 26-year-old Brazilian who took a £50 a week cleaning job at the Kensington International Hotel after spending the 2000 American dollars he had brought with him; he had started the job a week after arriving in Britain to study English.)

(*The Times*, 19 June 1987)

This is not only a comment on the circumstances described but also on the daily ideological stance of Rupert Murdoch's News International's paper.

16. In the European Community market, companies diversify across the continent. European Silicon Structures, an international computer-chip company, is incorporated in Luxemburg, has its headquarters in Germany's Munich, its main factory in southern France, and its research centre in England.

(S. Greenhouse, 'Making Europe a mighty market', *New York Times*, 22 May 1988, p. 6)

SELECT BIBLIOGRAPHY

ABC Guide to London (1888) London: Joseph Smith.

Abu-Lughod, J. (1976) 'Developments in North African urbanism. The process of decolonization', in B.J.L. Berry (ed.) *Urbanization and Counterurbanization*, Beverly Hills, CA and London: Sage, pp. 191–211.

Abu-Lughod, J. (1978) 'Dependent urbanism and decolonization: the Moroccan Case', *Arab Studies Quarterly* 1: 49–66.

Abu-Lughod, J. (1980) *Rabat. Urban Apartheid in Morocco*, Princeton, NJ: Princeton University Press.

Abu-Lughod, J. (1984) 'Culture, "modes of production" and the changing nature of cities in the Arab world', in J. Agnew, J. Mercer, and D. Sopher (eds) *The City in Cultural Context*, London: Allen & Unwin, pp. 44–117.

Alger, C.F. (1988) 'Perceiving, analyzing and coping with the global-local nexus', *International Social Science Journal*, 117: 14–32.

Allman, T.D. (1983) 'The city of the future', *Esquire*, February, pp. 39–47.

Anderson, P. (1987) 'The figures of descent', *New Left Review*, 161: 20–77.

Armstrong, W. and McGee, T.G. (1985) *Theatres of Accumulation. Studies in Asian and Latin American Urbanization*, London: Methuen.

Aslet, C. (1982) *The Last Country Houses*, New Haven, CT: Yale University Press.

Aspinall, A. (1914) *A Guide to the West Indies*, London: Duckworth.

Baedeker, K. (1889) *London and its Environs*, London: Dulau.

The Banker, (1970–present).

Barras, R. (1981) 'The causes of the London office boom', in R. Barras (ed.) *The Office Boom in London*, proceedings of the first CES London Conference, London: CES, pp. 9–14.

Barratt Brown, M. (1978) *The Economics of Imperialism*, Harmondsworth: Penguin.

Becker, P.G., Frieden, B., Schatz, S.P., and Sklar, R.L. (1987) *Post-Imperialism. International Capitalism and Development in the late Twentieth Century*, Boulder, CO and London: Rienner.

Blackaby, F.T. (1979) *De-industrialisation*, London: Heinemann.

Bohn, H.G. (1854) *The Pictorial Handbook of London*, London: Bohn.

Braudel, F. (1984) *The Perspective of the World*, London: Fontana.

Briggs, A. (1963) 'London: the World City', in ibid, *Victorian Cities*, London: Odhams, pp. 321-72.

Brimson, P. (1979) *Islington's Multinationals*, London: Islington Economy Group.

British Council (1984) *Overseas Students in the United Kingdom 1983-4*, London: British Council.

British Council (1988) *Statistics of Overseas Students in the UK. 1985-86*, London: British Council.

Brown, G.G. (ed.) (1938) *The South and East African Year Book and Guide for 1936*, edited for the Union-Castle Mail Steamship Company Ltd, London: Sampson Low, Marston & Co. (42nd issue).

Browning, H. and Roberts, B. (1980) 'Urbanisation, sectoral transformation and the utilization of labour in Latin America', *Comparative Urban Research*, 8 (1): 86-104.

Brunn, S.D. and Williams, J.F. (1983) *Cities of the World: World Regional Urban Development*, New York: Harper and Row.

Cain, P.J. and Hopkins, A.G. (1980) 'The political economy of British expansion overseas, 1750-1914', *Economic History Review* 33: 463-90.

Cannadine, D. and Reeder, D. (1982) *Exploring the Urban Past. Essays in Urban History by H.J. Dyos*, Cambridge: Cambridge University Press.

Canning Town Community Development Project (1977) 'Canning Town's economy, 1846-1976', in *Growth and Decline*, London: Canning Town CDP.

Carr, M.C. (1982) 'The development and character of a metropolitan suburb: Bexley, Kent', in F.M.L. Thompson (ed.) *The Rise of Suburbia*, Leicester: Leicester University Press, pp. 211-69.

Carstairs, G.M. (1968) *The Twice-Born*, London: Allen and Unwin.

Castells, M. (1977) *The Urban Question*, London: Edward Arnold (first published as *La Question Urbaine*, Paris, Francis Maspero (1972).

Castle, S. and Kossack, G. (1973) *Immigrant Workers and the Class Structure*, Oxford: Oxford University Press.

Chaichian, M.A. (1988) 'The effects of world capitalist economy on urbanization in Egypt, 1800-1970', *International Journal of Middle Eastern Studies* 20: 23-43.

Chase-Dunn, C. (1985) 'The system of world cities, AD 800-1975', in M. Timberlake (ed.) *Urbanization in the World-Economy*, London: Academic Press.

Cheshire, P. and Hay, D. (1986) 'The development of the European urban system 1971-1981', in H-J. Ewers, J.B. Goddard, and H. Matzerath (eds) *The Future of the Metropolis*, Berlin and New York: de Gruyter.

Christopher, A.J. (1988) *The British Empire at its Zenith*, London: Croom Helm.

Clarke, W.M. (1969) *The City and the World Economy*, Harmondsworth: Pelican.

Clout, H. and Wood, P. (1986) *London. Problems of Change*, London: Longman.

Cohen, R.B. (1981) 'The new international division of labor, multinational corporations and urban hierarchy', in M. Dear and A.J. Scott (eds) *Urbanization and Urban Planning in Capitalist Society*, London: Methuen, pp. 287–315.

Cohen, R. (1987) *The New Helots. Migrants in the International Division of Labour*, Aldershot: Gower Publishing.

Consulting Engineer's Who's Who and Year Book (1982–3), London: Municipal Publications.

Conway, J. (1985) *Capital Decay. An Analysis of London's Housing*, SHAC Report 7, London: SHAC.

Cooke, P. (1986a) 'The changing urban and regional system in the UK', *Regional Studies* 20: 243–51.

Cooke, P. (ed.) (1986b) *Global Restructuring, Local Response*, Redhill, Surrey: Schools Publishing Co.

Cooke, P. (1988) 'Modernity, postmodernity and the city', *Theory, Culture and Society* 5 (2–3): 475–93.

Cooke, P. (ed.) (1989) *Localities*, London: Hutchinson (forthcoming).

Cooke, P. and Thrift, N. (eds) (1989) *Captive Britain*, Cambridge: Cambridge University Press.

Cornforth, J. (1974) *Country Houses in Britain. Can They Survive?* London: *Country Life* for the British Tourist Authority.

Council of Europe (Council for Cultural Co-operation) (1986) List of non-recognized institutions of higher education, Directorate of Education, Culture and Sport Secretariat, Documentation Section, Strasbourg.

Counter Information Services (CIS) (1983) *Banking on the City*, London: CIS.

Cowan, H. (1987) *The Cheltenham Ambience. Reading the Built Environment*, Cheltenham Locality Study, ESRC 'Changing Urban and Regional System in the UK', Initiative, Working Paper 2, Gloucester College of Arts and Technology, Gloucester.

Cruickshank, D. (1987) 'India's legacy', *Landscape* 2: 74–8.

Damesick, P. (1980) 'The inner city economy in industrial and post-industrial London', *London Journal* 6 (1): 23–35.

Daniels, P.W. (1986) 'The geography of services', *Progress in Human Geography.*

Darley, G. (1985) 'Existential cities', in R. Fermor-Hesketh (ed.) *Architecture of the British Empire*, London: Weidenfeld & Nicholson, pp. 74–103.

Darlow, C. (1986) *The London Property Market in AD 2000*, London: Spon.

Davis, H. (1934) *The South American Handbook*, London: Trade and Travel Publications Ltd.

Davis, M. (1985) 'Urban renaissance and the spirit of postmodernism', *New Left Review* 151: 106–14.

Dear, M.J. (1986) 'Postmodernism and planning', *Society and Space* 4: 367–84.

Dear, M. and Scott, A.J. (eds) (1981) *Urbanization and Urban Planning in Capitalist Society*, London: Methuen.

Dennis, R. (1984) *English Industrial Cities of the Nineteenth Century: a Social Geography*, Cambridge: Polity Press, pp. 309–24.

Dickens, P. (1986) *Global Shift. Industrial Change in a Turbulent World*, London: Harper & Row.

Dolphin, P., Grant, E., and Lewis, E. (1981) *The London Region: An Annotated Geographical Bibliography*, London: Unwin.

Douglas, M. (1978) *Purity and Danger*, London: Routledge & Kegan Paul.

Drakakis-Smith, D. (ed.) (1986) *Urbanization in the Developing World*, London: Croom Helm.

Drakakis-Smith, D. (1987) *The Third World City*, New York: Methuen.

Drummond, I. (1981) 'Britain and the world economy, 1900-45', in R. Floud and D. McCloskey (eds) *The Economic History of Britain since 1700, part 2, 1860-1970s*, Cambridge: Cambridge University Press, pp. 286-308.

Duffy, F. (1983) 'Information technology, organizations and the office', *Orbit* (Office Research into Buildings and Information Technology), London: DEGW.

Duffy, F. (1987) 'How the expansion of financial services is influencing office accommodation', paper presented to the British Sociological Association Study Group on Sociology and Environment meeting, 'The built environment and global restructuring', 21 November (typescript).

Dunning, J.H. (1983) 'Changes in the level and structure of international production: the last one hundred years', in M. Casson (ed.) *The Growth of International Business*, London: Allen & Unwin, pp. 84-139.

Dunning, J.H. and Norman, G. (1979) 'Factors influencing the location of offices of multinational enterprises', *Location of Offices Research Paper*, vol. 8, London: Economists Advisory Group.

Dunning, J.H. and Norman, G. (1983) 'The theory of multinational enterprise: an application to multinational office location', *Environment and Planning A* 15: 675-92.

Dunning, J.H. and Norman, G. (1987) 'The location choice of international companies', *Environment and Planning A*, 19 (5): 613-31.

Dunning, J.H. and Pearce, R.D. (1981) *The World's Largest Industrial Enterprises*, London: Gower.

Dyer, M. (1982) 'City's conservation areas doubled', *Chartered Surveyor*, February, pp. 390-1.

Dyos, J. and Wolff, M. (1973) *The Victorian City*, 2 vols, London: Routledge & Kegan Paul.

Economist (1986) 'Capital of capital', 11 October, pp. 13-14.

Edelstein, M. (1981) 'Foreign investment in empire', in R. Floud and D. McCluskey (eds) *The Economic History of Britain since 1700, part 2, 1860-1970s*, Cambridge: Cambridge University Press, pp. 70-98.

Elliott, M. (1986) *Heartbeat London*, London: Firethorn Press.

Ellsworth, J.B., Mundheim, R.H., and Hawes, D.W. (eds) (1981) 'Proceedings of the Conference on the Internationalization of the Capital Markets', *Journal of Comparative Corporate Law and Securities Regulation* 3: 3-4.

Estate Times (1985) *Brave New World? City of London Office Property and the Financial Services Revolution*, London: Estate Times.

Evans, A. (1973) 'The location of headquarters of industrial companies', *Urban Studies* 10: 387–95.

Evans, A. and Eversley, D. (eds) (1980) *The Inner City: Employment and Industry*, London: Heinemann.

Evers, H.D. (1984) 'Urban land ownership, ethnicity and class in South-East Asian cities', *International Journal of Urban and Regional Research* 8 (4): 481–96.

Eversley, D. (1980) 'A planner's perspective', in A. Evans and D. Eversley (eds) *The Inner City: Employment and Industry*, London: Heinemann, pp. 460–78.

Ewars, H.J., Goddard, J.B. and Matzerath, H. (eds) (1986) *The Future of the Metropolis*, Berlin and New York: Walter de Gruyter.

Falk, N. (1981) 'London's docklands: a tale of two cities', *London Journal* 7: 65–80.

Feagin, J.R. (1985) 'The global context of metropolitan growth: Houston and the oil industry', *American Journal of Sociology* 90 (6): 1204–30.

Feagin, J.R. and Smith, M.P. (1987) 'Cities and the new international division of labour. An overview', in M.P. Smith and J.R. Feagin (eds) *The Capitalist City*, Oxford: Blackwell.

Fetter, B. (1976) *The Creation of Elizabethville, 1910-40*, Stanford: Hoover Institute Press.

Floud, R. and McCloskey, D. (1981) *The Economic History of Britain since 1700, part 2, 1860-1970s*, Cambridge: Cambridge University Press.

Forbes, D. and Thrift, N.J. (eds) (1987) *The Socialist Third World. Urban Development and Territorial Planning*, London: Blackwell.

Foster, J. (1977) *Class Struggle and the Industrial Revolution*, London: Methuen.

Fothergill, S. and Gudgin, G. (1982) *Urban and Regional Employment Change in the UK*, London: Heinemann.

Friedmann, J. (1986) 'The World City Hypothesis', *Development and Change* 17 (1): 69–83.

Friedmann, J. and Wolff, G. (1982) 'World City formation: an agenda for research and action', *International Journal of Urban and Regional Research* 6: 309–44.

Frobel, F., Heinrichs, J., and Kreye, O. (1980) *The New International Division of Labour*, Cambridge: Cambridge University Press.

Gamble, A. (1985) *Britain in Decline*, London: Macmillan.

Garside, P.L. (1984) 'East and West: the world of London, 1890–1940', in A. Sutcliffe (ed.) *Metropolis, 1890-1940*, London: Mansell.

Girouard, M. (1971) *The Victorian Country House*, London: Country Life.

Glickman, N.J. (1987) 'Cities and the international division of labour', in M.P. Smith and J.R. Feagin (eds) *The Capitalist City*, Oxford: Blackwell, pp. 66–86.

Goldfrank, W.L. (ed.) (1979) *The World System of Capitalism: Past and Present*, Beverly Hills and London: Sage.

Golland, R. (1980) 'Inner city industry: a literature review', Greater London Council, Research Documents Guide, vol. 3.

Goodhart, D. and Grant, G. (1986) 'The internationalization of capital', *New Statesman*, 24 October, pp. 6–9.

Greater London Council (GLC) (1981a) Planning Committee, report on Review of Office Policy in Central London, GLC.

Greater London Council (1981b) 'London: economic trends and issues, 1981', reviews and studies series, vol. 10.

Greater London Council (1983a) 'Paid to think: professional workers in London', Economic Policy Group, strategy document no. 25.

Greater London Council (1983b) Census results for Greater London and the London Boroughs: small area statistics and historical comparisons, statistical series, 19.

Greater London Council (1985) *The London Industrial Strategy*, London: GLC.

Greater London Council (1986) *London Facts and Figures*, GLC Intelligence Unit.

Gregory, D. and Urry, J. (1985) *Social Relations and Spatial Structures*, New York: Macmillan.

Griffith, B., Payner, G., and Mohan, J. (1985) *Commercial Medicine in London*, London: Greater London Council Industry and Employment Branch.

Gripaios, P. (1977) 'Industrial decline in London: an examination of causes', *Urban Studies*, 14: 181-9.

Gripaios, P. (1980) 'Economic decline in south-east London, 1980', in A. Evans and D. Eversley (eds) *The Inner City: Employment and Industry*, London: Heinemann, pp. 65-77.

Grunwald, J. and Flamm, K. (1985) *The Global Factory*, Washington: Brookings Inst.

Gulf Leisure Investments (1986).

Gutman, R. (1988) *Architectural Practice*, New York: Princeton.

Hall, J.M., Griffiths, G., Eyles, J., and Darby, M. (1976) 'Rebuilding the London Docklands', *London Journal* 2 (2): 226-81.

Hall, P. (1984) *The World Cities*, London: Weidenfeld & Nicolson (3rd edn; 1st edn, 1966).

Hall, P. (1987) 'The anatomy of job creation: nations, regions and cities in the 1960s and 1970s', *Regional Studies* 21 (2): 95-106.

Hamilton, A. (1986) *The Financial Revolution, The Big Bang World Wide*, Harmondsworth: Penguin.

Hamnett, C. (1986) 'The changing socio-economic structure of London and the South-East, 1961-81', *Regional Studies*, 20 (5): 391-406.

Hamnett, C. (1987) 'A tale of two cities: sociotenurial polarization in London and the South-East, 1966-81', *Environment and Planning A*, 19: 537-56.

Hamnett, C. and Randolph, B. (1982) 'How far will London's population fall?', *London Journal* 8 (1): 95-100.

Harley, C.K. and McCloskey, D.N. (1981) 'Foreign trade: competition and the expanding international economy', in R. Floud and D. McCloskey (eds) *The Economic History of Britain since 1700, part 2, 1860-1970s*, Cambridge: Cambridge University Press, pp. 50-70.

Harloe, M. (1987) 'Editorial introduction', *International Journal of Urban and Regional Research* 11: 1.

Harris, N. (1987) *The End of the Third World*, London: Taurus.

Hart, G. (1981) *A History of Cheltenham*, Gloucester: Alan Sutton Publishing.

Harvey, D. (1973) *Social Justice and the City*, London: Edward Arnold.

Harvey, D. (1987) 'Flexible accumulation through urbanization: reflections on post-modernism in the American city', paper presented to the Sixth Urban Change and Conflict Conference, University of Kent, September.

Henderson, J. and Castells, M. (1987) *Global Restructuring and Territorial Development*, London and New York: Sage.

Heenan, D.A. (1977) 'Global cities of tomorrow', *Harvard Business Review* 55 (3): 79-92.

Hill, R.C. (1984) 'Urban political economy', in M.P. Smith (ed.) *Cities in Transformation*, Beverly Hills, Calif. and London: Sage, pp. 123-38.

Hill, R.C. and Feagin, J.R. (1987) 'Detroit and Houston; two cities in global perspective', in M.P. Smith and J.R. Feagin (eds) *The Capitalist City*, Oxford: Blackwell, pp. 155-77.

Hillier Parker (1985) *International Property Bulletin*, Hillier Parker International Department, 77 Grosvenor Street, London W1A 2BT.

Hillier Parker (1987) *International Property Bulletin*, Hillier Parker International Department, 77 Grosvenor Street, London W1A 2BT.

Hoare, G. (1975) 'Foreign firms and air transport: the geographical effect of Heathrow airport', *Regional Studies* 9: 349-67.

Hobsbawm, E.J. (1975) *The Age of Capital*, London: Weidenfeld & Nicolson.

Hopkins, A.G. (1980) 'Property rights and empire building: the British annexation of Lagos, 1861', *Journal of Economic History* 40: 777-98.

Horvath, R.V. (1969) 'In search of a theory of urbanization: notes on the colonial city', *East Lakes Geographer* 5: 68-82.

Hudson, R. and Williams, A. (1986) *The United Kingdom*, London: Harper & Row.

Ingham, G. (1984) *Capitalism Divided*, London: Macmillan.

Institute of International Education (1986a) *Open Doors*, New York: IIE (1976-86).

Institute of International Education (1986a) *Study in the UK and Ireland*, New York: IIE.

Irving, R.G. (1983) *Indian Summer, Lutyens, Baker & Imperial Delhi*, New Haven: Yale University Press.

Jackson, A.A. (1973) *Semi-detached London*, London: Allen & Unwin.

Jahn, M. (1983) 'Suburban developments in outer west London, 1850-1900', in P.M.L. Thompson (ed.) *The Rise of Suburbia*, Leicester: Leicester University Press, pp. 93-156.

Jameson, F. (1984) 'Postmodernism, or the cultural logic of late capitalism', *New Left Review* 148: 53-92.

Jenkins, R. (1971) *Exploitation. The World Power Structure and the Inequality of Nations*, London: Paladin.

Johnson, J.H. and Pooley, C. (1982) *The Structure of Nineteenth Century Cities*, London: Croom Helm.

Jones Lang Wootton (1980) *Offices in the City of London*, 'A special report', London: Jones Lang Wootton.

Jones Lang Wootton (1984) *Foreign Investment in US Real Estate, 1979–83*, Jones Lang Wootton Investment Research, September.

Jones Lang Wootton (1986) *Central London Office Research Mid-1986*, London: Jones Lang Wootton.

Jones Lang Wootton (1987) *Office Development in Greater London outside the Central Area*, JLW Annual Survey, London: Jones Lang Wootton.

Kentor, J. (1985) 'Economic development and the world division of labour', in M. Timberlake (ed.) *Urbanization in the World-Economy*, London: Academic Press, pp. 25–40.

King, A.D. (1962) 'As she is spoke', *Economist*, 15 September, no. 1013 (published under 'A correspondent').

King, A.D. (1976) *Colonial Urban Development: Culture, Social Power and Environment*, London: Routledge & Kegan Paul.

King, A.D. (1983) ' "The world economy is everywhere"; urban history and the world system', *Urban History Yearbook*, Leicester: Leicester University Press.

King, A.D. (1984a) *The Bungalow: the Production of a Global Culture*, London: Routledge & Kegan Paul.

King, A.D. (1984b) 'Capital city: physical and social aspects of London's role in the world-economy', UCLA project on World Cities in Formation (unpublished).

King, A.D. (1986a) 'Margins, peripheries and divisions of labour: UK urbanism and the world-economy', in D. Hardy (ed.) *On the Margins*, Middlesex Polytechnic Planning Papers.

King, A.D. (1987a) 'Making a market in meaning: the national and international reorganization of domestic architecture', paper for BSA Study Group on Sociology and Environment, 'The built environment and global restructuring', London, 21 November (unpublished).

King, A.D. (1987b) 'Cultural production and reproduction: the political economy of societies and their built environment', in D. Cantor, M. Krampen, and D. Stea (eds) *Ethnoscapes. Transcultural Studies in Action and Place*, London: Gower.

King, A.D. (1988) 'Beirut: internal and external ecologies', in Liazu M. Claude (ed.) *Etat, Ville et Mouvements sociaux au Maghreb et au Moyen Orient*, Paris: CNRS et ESRC (forthcoming).

King, A.D. (1989a), 'Colonialism, urbanism, and the capitalist world-economy: an introduction', *International Journal of Urban and Regional Research* 13 (1): 1–18.

King, A.D. (1989b) *Urbanism, Colonialism and the World-Economy*, London: Routledge.

Kira, A. (1976) *The Bathroom*, Harmondsworth, Penguin Books.

Knight, D.R.W., Tsapatsaris, A., and Jaroszek, J. (1977) 'The structure of employment in Greater London, 1961–81', Greater London Council Research Memorandum.

Knox, P. (1987) 'The social production of the built environment: architects, architecture and the post-modern city', *Progress in Human Geography* 11 (3): 354–77.

Korff, R. (1987) 'The world city hypothesis: a critique', *Development and Change* 18 (3): 483–92.

Kowarcik, L. and Campanario, M. (1986) 'S"ao Paulo: the price of world city status', *Development and Change* 17 (1): 154–74.

Kumar, K. (ed.) (1980) *Transnational Enterprises. Their Impact on Third World Societies and Culture*, Boulder, CO: Westview Press.

Langton, R.S. *et al.* (1981) *Development in London 1967–80*, London: Bernard Thorpe and Partners.

Lapping, A. (1977) 'London's burning! London's burning!', a survey, *Economist*, 1 January, 17–38.

Lee, T.R. (1978) *Race and Residence*, Oxford: Oxford University Press.

Lemon, B. (1987) 'The geography of the 'Big Bang': London's office building boom', *Geography*, 18: 56–7.

Light, I. (1988) 'Los Angeles', in M. Dogan and J.D. Kasarda (eds) *The Metropolis Era. Volume 2. Mega-Cities*, Beverly Hills and London: Sage, pp. 56–96.

Livingstone, K. (1982) 'Leader's Report to the Council', Greater London Council Meeting, 12 October.

Lomas, G. (1978) 'Inner London's future: studies and policies'; *London Journal* 14 (1): 95–105.

London Visitor and Convention Bureau (1985) *London Tourism Statistics, 1985*, LVCB, 26 Grosvenor Gardens, Victoria, London SW1

London Visitor and Convention Bureau (1987) 'Recent trends in tourism: visitors to London', LVCB, 26 Grosvenor Gardens, Victoria, London SW1.

London Tourist Board (1982) *Tourism in Greater London*, an assessment of current needs, London: London Tourist Board.

Love, J.H. and McNicoll (1988) 'The regional impact of overseas students in the UK: a case of three Scottish universities', *Regional Studies* 22 (1): 11–18.

Luff, D. (1982) 'City offices: eastward ho!', *Chartered Surveyor* 114 (7): 388–91.

Mandel, E. (1978) *Late Capitalism*, London: Verso.

Manpower Services Commission (MSC) (1982) *London Employment*, London: HMSO.

Marder, K.B. and Alderson, L.P. (1982) *Economic Society*, Oxford: Oxford University Press.

Massey, D. (1984) *Spatial Divisions of Labour*, London: Macmillan.

Massey, D. (1986a) 'The legacy lingers on: the impact of Britain's historical international role on its internal geography', in R. Martin and R. Rowthorn (eds) *Deindustrialization and the British Economy*, London: Macmillan.

Massey, D. (1986b) 'The international division of labour and local economic strategies: thoughts from London and Managua', paper presented to the ESRC Conference on 'Localities in an International Economy', UWIST, Cardiff, September.

Massey, D. (1988) 'A new class of geography', *Marxism Today*, May, pp. 12–15.

Massey, D. and Meegan, R. (1980) 'Industrial restructuring versus the cities', in A. Evans and D. Eversley (eds) *The Inner City: Employment and Industry*, London: Heinemann, pp. 78–107. (Previously in *Urban Studies* 15: 273–88.)

Massey, D. and Meegan, R. (1982) *The Anatomy of Job Loss*, London: Methuen.

McAuley, P. (1987) *Guide to Ethnic London*, London: Michael Haag.

McCrum, R., Cran, W., and MacNeil, R. (1986) *The Story of English*, London: Faber & Faber.

McGee, T.G. (1967) *The South East Asian City*, London: Bell.

McGee, T. (1986) 'Circuits and networks of capital: the internationalisation of the world economy and national urbanization', in D. Drakakis–Smith, (ed.) *Urbanization in the Developing World*, London: Croom Helm.

McRae, H. and Cairncross, F. (1985) *Capital City*, London: Methuen.

Meyer, D.R. (1986) 'The world system of cities: relations between international financial metropolises and South American cities', *Social Forces* 64 (3): 553–81.

Mitchell's Standard Guide to Buenos Aires (1910), London: T. Werner Laurie.

Mitter, S. (1986a) *Common Fate, Common Bond. Women in the Global Economy*, London: Pluto Press.

Mitter, S. (1986b) 'Industrial restructuring and manufacturing homework: immigrant women in the UK clothing industry', *Capital and Class* 27, Winter, pp. 37–80.

Mogg, E. (1841) *Mogg's New Picture of London or Strangers Guide to the British Metropolis*, London: E. Mogg.

Morgan, W.T.W. (1961) 'The two office districts of central London', *Journal of the Town Planning Institute* 47: 161–5.

Morgan, W.T.W. (1962) 'The geographical concentration of big business in Great Britain', *Town and Country Planning* 30 (3): 122–4.

Morris, E.K. (1978) 'Symbols of empire: architectural style and the government offices competition', *Journal of Architectural Education* November: 8–14.

Moulaert, F.G. and Salinas, P.W. (1983) *Regional Analysis and the New International Division of Labour*, The Hague and London: Kluwer Nijhoff.

Murray, R. (1985) 'London and the GLC: restructuring the capital of capital', *IDS Bulletin* 16 (1): 47–55.

National Westminster Bank (1982) *Annual Report and Accounts*, London: N.W.B.

National Westminster Bank (1983) *Annual Report and Accounts*, London: N.W.B.

Noyelle, T. (1983) 'The rise of advanced services. Some implications for economic development in US cities', *American Planners Association Journal*, Summer, pp. 280–70.

Noyelle, T. (1986) 'The international services economy', paper presented to the ESRC Conference on 'Localities in an International Economy', UWIST, Cardiff, September.

Noyelle, T.J. and Stanback, T.M. (1984) *The Economic Transformation of American Cities*, Totowa, NJ: Rowman and Allanheld.

182

Office of Population Censuses and Surveys (OPCS) (1973) *Census 1971,*
    County Report: Greater London, part 1, London: HMSO.
Office of Population Censuses and Surveys (OPCS) (1982) *Census 1981,*
    County Report: Greater London, part 1, London: HMSO.
Page, S. (1987) 'The London Docklands: redevelopment schemes in the
    1980s', *Geography* 72: 59–63.
Pahl, R.E. (1970) *Patterns of Urban Life,* London: Longman.
Pahl, R.E., Flynn, R., and Buck, N.H. (1983) *Structures and Processes of
    Urban Life,* London: Longman.
Pang, E.S. (1983) 'Buenos Aires and the Argentine economy in world
    perspective, 1776–1930', *Journal of Urban History,* 9 (3): 365–82.
Peach, C., Robinson, V., and Smith, S. (1980) *Ethnic Segregation in Cities,*
    London: Croom Helm.
Perry, D.C. (1987) 'The politics of dependency in de-industrializing
    America. The case of Buffalo, New York', in M.P. Smith and J.R.
    Feagin (eds) *The Capitalist City,* Oxford: Blackwell, pp. 113–37.
Petras, E. (1981) 'The global labor market in the modern world-economy',
    in M.M. Kritz, C.B. Keely, and S.M. Tomasi (eds) *Global Trends in
    Migration: Theory and Research on International Population Movements,* New
    York: Centre for Migration Studies of New York, Inc., pp. 44–63.
Porter, B. (1979) *The Lion's Share. A Short History of British Imperialism,*
    London: Longman.
Portes, A. and Walton, J. (1981) *Labor, Class and the International System,*
    London: Academic Press.
Pred, A. (1980) *City-Systems in Advanced Economies. Past Growth, Present
    Processes and Future Development Options,* New York: Wiley.
Property International ('The magazine covering leisure and investment real
    estate around the world') (1985), vols 1 and 2, Falcon Publishing
    Europe Ltd, 27–29 Queen Anne St, London W1V.
Pugh, E. (n.d.) *The City of the World,* London: Nelson.
Rabinow, P. (1989) 'Governing Morocco: modernity and difference',
    International Journal of Urban and Regional Research 13 (1).
Rayfield, J.R. (1974) 'Theories of urbanization and the colonial city in
    West Africa', *Africa* 44: 163–85.
Rex, J. (1973) *Race, Colonialism and the City,* London: Routledge & Kegan Paul.
Ribeiro, A. (1989) 'The creation of a real estate market in Rio de
    Janeiro, 1870–1930', *International Journal of Urban and Regional Research*
    13 (1).
Rimmer, P.J. (1986) 'Japan's world cities: Tokyo, Osaka, Nagoya or
    Tokaido Megalopolis?' *Development and Change* 17 (1): 120–50.
Roberts, B. (1978a) *Cities of Peasants,* London: Arnold.
Roberts, B. (1978b) 'Comparative perspectives on urbanization', in D.
    Street and associates, *Handbook of Contemporary Urban Life,* San
    Francisco and London: Jossey-Bass, pp. 592–627.
Robertson, R. (1985) 'The sacred and the world system', in P.E.
    Hammond (ed.) *The Sacred in a Secular Age. Towards Revision in the
    Scientific Study of Religion,* Los Angeles and London: University of
    California Press.

Robertson, R. (1987) 'Globalization theory and civilizational analysis', *Comparative Civilizations Review* 17: 20–30.

Robertson, R. (1988) 'The sociological significance of culture: some general considerations', *Theory, Culture and Society* 5: 3–24.

Robertson, R. (1989) 'Globality, global culture and images of world order', in H. Haverkamp and H. Smelser (eds) *Social Change and Modernity*, Berkeley, CA: University of California Press (in press).

Robertson, R. and Lechner, F. (1985) 'Modernization, globalization and the problem of culture in world-systems theory', *Theory, Culture and Society* 2 (3): 103–118.

Robson, B. (1973) *Urban Growth. An Approach*, London: Methuen.

Ross, R. and Trachte, K. (1983) 'Global cities and global classes. The peripheralization of labor in New York City', *Review* 6 (3): 393–431.

Royal Institute of British Architects (RIBA) (1980) *Directory of International Practices, 1980–1*, London: RIBA.

Royal Institute of British Architects (1986) *RIBA International Directory of Practice*, London: RIBA.

Salinas, P.W. (1983) 'Mode of production and spatial organization in Peru', in F.G. Moulaert and P.W. Salinas (eds) *Regional Analysis and the New International Division of Labour*, The Hague and London: Kluwer-Nijhoff, pp. 79–96.

Sassen-Koob, S. (1984) 'The new labor demand in global cities', in M.P. Smith (ed.) *Cities in Transformation*, Beverly Hills and London: Sage, pp. 139–72.

Sassen-Koob, S. (1985) 'Capital mobility and labor migration: their expression in core cities', in M. Timberlake (ed.) *Urbanization in the World-Economy*, London: Academic Press, pp. 231–68.

Sassen-Koob, S. (1986) 'New York City: economic restructuring and immigration', *Development and Change* 17 (1): 85–119.

Sassen-Koob, S. (1987) 'Growth and informalization at the core. A preliminary report on New York City', in M.P. Smith and J.R. Feagin (eds) *The Capitalist City*, Oxford: Blackwell, pp. 138–54.

Saueressig-Schreuder, Y. (1986) 'The impact of British colonial rule on the urban hierarchy of Burma', *Review* 10 (2): 245–77.

Saunders, P. (1981) *Social Theory and the Urban Question*, London: Hutchinson.

Savills Magazine, Savills, 20 Grosvenor Hill, London W1.

Scott, B. (1867) *A Statistical Vindication of the City of London*, London: Longmans, Green, Reader, and Dyer.

Service, A. (1978) *London, 1900*, London: Crosby.

Seymour, H. (1986) 'The international construction industry', in Bartlett International Summer School, *The Production of the Built Environment*, 7, Bartlett School of Architecture and Planning, University College, London.

Shawcross, T. and Fletcher, K. (1987) 'How crime is organized in London', *The Illustrated London News*, pp. 33–39.

Shelter (1985) 'Homelessness', *Housing Factsheet 1*, 157 Waterloo Road, London SE1.

Shelter, (1987) *English Homelessness Statistics*, 3rd Quarter, 157 Waterloo Road, London SE1.

Simmons, C. and Kirk, R. (1981) 'Lancashire and the equipping of the Indian cotton mills: a study of textile machinery supply, 1854–1939', *Proceedings of the Seventh European Conference on Modern South Asian Studies*, School of Oriental and African Studies, London, July.

Simon, D. (1989) 'Colonial cities, post-colonial Africa and the world-economy: some pertinent issues and questions', *International Journal of Urban and Regional Research* 13 (1).

Singh, A. (1977) 'UK industry and the world-economy: a case of deindustrialization?' *Cambridge Journal of Economics* 1: 113–36.

Slater, D. (1980) 'Towards a political economy of urbanization in peripheral capitalist societies. Problems of theory and method with illustrations from Latin America', *International Journal of Urban and Regional Research* 4 (3): 27–51.

Slater, D. (1986) 'Capitalism and urbanization at the periphery', in D. Drakakis-Smith (ed.) *Urbanization in the Developing World*, London: Croom Helm.

Smith, M. (ed.) (1983) 'Structuralist urban theory: a symposium', *Comparative Urban Research Research* 9: 5–70.

Smith, M.P. (1980) *The City and Social Theory*, Oxford: Oxford University Press.

Smith, M.P. and Feagin, J.R. (1987) *The Capitalist City*, Oxford: Blackwell.

Soja, E. (1986) 'Modernity and locality: internationalization in Greater Los Angeles', paper presented to the ESRC conference on 'Localities in an International Economy', UWIST, Cardiff, September.

Soja, E.W. and Weaver, C.E. (1976) 'Urbanization and underdevelopment in East Africa', in B.J.L. Berry (ed.) *Urbanization and Counter-Urbanization*, Beverly Hills, Calif. and London: Sage, pp. 233–66.

Soja, E., Heskin, A.D., and Cenzatti, M. (1985) 'Los Angeles: through the kaleidoscope of urban restructuring', UCLA Graduate School of Architectural and Urban Planning.

Soja, E., Morales, R., and Wolff, G. (1983) 'Urban restructuring: an analysis of social and spatial change in Los Angeles', *Economic Geography* 59: 195–230.

Southwark Trade Council and Roberts, J. (1976) 'Employment in Southwark. A strategy for the future', Southwark Trades Council.

Steinberg, F. (1984) 'Town planning and the neo-colonial modernization of Colombo', *International Journal of Urban and Regional Research* 8 (4): 530–48.

Sternlieb, G. and Hughes, J.W. (1988) 'New York City', in M. Dogan and J.D. Kasarda (eds) *The Metropolis Era, Volume 2. Mega-Cities*, Beverly Hills and London: Sage, pp. 20–55.

Tabb, W. and Sawers, L. (eds) (1984) *Marx and the Metropolis*, New York: Oxford University Press.

The Times 1000, 1986–7, London: Times Publishing.

Thrift, N.J. (1985) 'Research policy and review 1. Taking the rest of the world seriously? The state of British urban and regional research in a time of economic crisis', *Environment and Planning A* 17: 7–24.

Thrift, N.J. (1986a) 'The internationalization of producer services and the integration of the Pacific Basin property market', in M.J. Taylor and N.J. Thrift (eds) *Multinationals and the Restructuring of the World-Economy*, London: Croom Helm.

Thrift, N.J. (1986b) 'The geography of international economic disorder', in R.J. Johnston and P.J. Taylor (eds) *A World in Crisis? Geographical Perspectives*, Oxford: Blackwell, pp. 12–67.

Thrift, N.J. (1987a) 'The fixers: the urban geography of international commercial capital', in J. Henderson and M. Castells (eds) *Global Restructuring and Territorial Development*, London: Sage.

Thrift, N.J. (1987b) 'Serious money. Capitalism, class, consumption and culture in late twentieth century Britain', paper presented to the IBG Conference on 'New Directions in Cultural Geography', London, September.

Thrift, N., Leyshon, A. and Daniels, P. (1987) 'Sexy greedy. The new international financial system, the City of London and the South East of England', paper for the Sixth Urban Change and Conflict Conference, University of Kent, September 1987 (forthcoming, Thrift, N. and Leyshon, A.L. (1989) *Making Money. The City of London and Social Power in Britain*, London: Routledge).

Timberlake, M. (ed.) (1985) *Urbanization in the World-Economy*, London: Academic Press.

Timberlake, M. (1987) 'World-system theory and the study of comparative urbanization', in M.P. Smith and J.R. Feagin (eds) *The Capitalist City*, Oxford: Blackwell, pp. 37–65.

Tolchin, M. and Tolchin, S. (1988) *Buying into America. How far Money is Changing the Face of the Nation*, New York: Times Books.

Townsend, P. (1987) *Poverty and Labour in London*, London: Low Pay Unit.

Trachte, K. and Ross, R. (1985) 'The crisis of Detroit and the emergence of global capitalism', *International Journal of Urban and Regional Research* 9 (2): 186–217.

United Nations (1986) *Demographic Yearbook, 1984*, New York: United Nations.

Urry, J. (1987) 'On the waterfront', *New Society*, 14 August, pp. 17–19.

Wald, M.L. (1988) 'Foreign investors take more active roles', *New York Times. Real Estate Report on Commercial Property*, 15 May, pp. 5–12.

Wallerstein, I. (1974) *The Modern World System*, London: Academic Press.

Wallerstein, I. (1979) *The Capitalist World-Economy*, Cambridge: Cambridge University Press.

Wallerstein, I. (1983) *Historical Capitalism*, London: Verso.

Wallerstein, I. (1984) *The Politics of the World-Economy*, Cambridge: Cambridge University Press.

Wallerstein, I. (1987) 'World systems analysis', in A. Giddens and J.H. Turner (eds) *Social Theory Today*, Cambridge: Polity Press, 309–24.

Walton, J. (1976) 'Political economy of world urban systems: directions for comparative research', in J. Walton and L. Massotti (eds) *The City in Comparative Perspective*, London and Beverly Hills: Sage, pp. 301–13.

Walton, J. (1982) 'The theory of peripheral urbanization', in N.L. Fainstein and S. Fainstein (eds) *Urban Policy under Capitalism*, Beverly Hills: Sage, pp. 118–34.

Walton, J. (1984) 'Culture and economy in shaping urban life: general issues and Latin American examples', in J.A. Agnew, J. Mercer, and D. Sopher (eds) *The City in Cultural Context*, London: Allen & Unwin, pp. 76–93.

Walton, J. (1985) *Capital and Labor in the Urbanized World*, London: New York, Sage.

Walton, J. (1987) 'Urban protest and the global political economy. The IMF riots', in M.P. Smith and J.R. Feagin (eds) *The Capitalist City*, Oxford: Blackwell, pp. 364–86.

Ward, Lock (1909) *London*, London: Ward, Lock.

Warren, B. (1980) *Imperialism. Pioneer of Capitalism*, London: Verso.

Weaver, C. and Richards, P. (1985) 'Planning Canada's role in the new global economy', *American Planners Association Journal*, Winter, pp. 43–51.

Weber, A.F. (1899) *The Growth of Cities in the Nineteenth Century*, Ithaca, New York: Cornell University Press (Reprint, Ithaca, 1963).

Weightman, G. and Humphries, S. (1984) *The Making of Modern London*, London: Sidgewick and Jackson.

Western, J. (1985) 'Undoing the colonial city', *Geographical Review* 75 (3): 335–57.

Wicks, M. (1987) 'The welfare gap', *New Society*, February.

Williams, P. and Smith, N. (1986) *Gentrification in the City*, London: Allen & Unwin.

Wolf, E. (1982) *Europe and the People without History*, Berkeley, CA: University of California Press.

Wolff-Phillips, L. (1987) 'Why "Third World"?: origin, definition and usage', *Third World Quarterly* 9 (4): 1311–27.

Woodruff, W. (1979) 'The emergence of an international economy, 1700–1914', in C.M. Cipolla (ed.) *The Fontana Economic History of Europe, The Emergence of Industrial Societies, 2*, London: Collins, pp. 656–37.

World Bank (1987) *World Development Report, 1987*, Oxford: Oxford University Press.

Wrigley, E. (1978) 'A simple model of London's importance in changing English society and economy, 1650–1750', in P. Abrams and E.A. Wright (eds) *Towns in Societies. Essays in Economic History and Historical Sociology*, Cambridge: Cambridge University Press, pp. 215–43.

Yannopolis, G.N. (1983) 'The growth of transnational banking', in N. Casson (ed.) *The Growth of International Business*, London: Allen & Unwin, pp. 236–57.

Young, K. and Mills, L. (1983) *Managing the Post-Industrial City*, London: Heinemann.

Zukin, S. (1988) 'The postmodern debate over urban form', *Theory, Culture, and Society* 5 (2–3): 431–46.

INDEX